1st Nano-IBCT Conference 2011

Radiation Damage of Biomolecular Systems: Nanoscale Insights into Ion Beam Cancer Therapy

Journal of Physics: Conference Series Volume 373

Caen, France
2-6 October 2011

Editors:

Bernd A Huber
Christiane Malot
Alicja Domaracka
Andrey V Solov'yov

ISBN: 978-1-62276-807-3
ISSN: 1742-6588

Printed from e-media with permission by:

Curran Associates, Inc.
57 Morehouse Lane
Red Hook, NY 12571

Printed by Curran Associates, Inc. (2013)

For permission requests, please contact the Institute of Physics at the address below.

Institute of Physics
Dirac House, Temple Back
Bristol BS1 6BE UK

Phone: 44 1 17 929 7481
Fax: 44 1 17 920 0979

techtracking@iop.org

Additional copies of this publication are available from:

Curran Associates, Inc.
57 Morehouse Lane
Red Hook, NY 12571 USA
Phone: 845-758-0400
Fax: 845-758-2634
Email: curran@proceedings.com
Web: www.proceedings.com

TABLE OF CONTENTS

Preface

The 1st Nano-IBCT Conference entitled 'Radiation Damage in Biomolecular Systems: Nanoscale Insights into Ion Beam Cancer Therapy' was held in Caen, France, in October 2011. The Meeting was organised in the framework of the COST Action MP1002 (Nano-IBCT) which was launched in December 2010 (http://fias.uni-frankfurt.de/nano-ibct). This action aims to promote the understanding of mechanisms and processes underlying the radiation damage of biomolecular systems at the molecular and nanoscopic level and to use the findings to improve the strategy of Ion Beam Cancer Therapy. In the hope of achieving this, participants from different disciplines were invited to represent the fields of physics, biology, medicine and chemistry, and also included those from industry and the operators of hadron therapy centres.

Ion beam therapy offers the possibility of excellent dose localization for treatment of malignant tumours, minimizing radiation damage in normal healthy tissue, while maximizing cell killing within the tumour. Several ion beam cancer therapy clinical centres are now operating in Europe and elsewhere. However, the full potential of such therapy can only be exploited by better understanding the physical, chemical and biological mechanisms that lead to cell death under ion irradiation. Considering a range of spatio-temporal scales, the proposed action therefore aims to combine the unique experimental and theoretical expertise available within Europe to acquire greater insight at the nanoscopic and molecular level into radiation damage induced by ion impact. Success in this endeavour will be both an important scientific breakthrough and give great impetus to the practical improvement of this innovative therapeutic technique. Ion therapy potentially provides an important advance in cancer therapy and the COST action MP1002 will be very significant in ensuring Europe's leadership in this field, providing the scientific background, required data and mechanistic insight which are indispensable for the optimization of this new therapy.

The conference gathered 115 participants originating from 28 countries and addressed a large number of highly relevant aspects concerning ion propagation in biological matter, the production of secondary particles along the ion tracks as electrons, holes and radicals, and their propagation in the biomolecular medium. In particular, the attack of DNA molecules and proteins by electrons and free radicals, the relative importance of direct and indirect damage processes as well as the role of the environment were discussed. Not only were fundamental mechanisms and processes elucidated, but radiobiological scale effects, multi-scale approaches and recent advances in the theoretical description of the underlying complex phenomena were also presented. Aspects linked to the energy deposition (LET), the characteristics of the Bragg peak and new techniques of dosimetry and radiolysis were highlighted. Furthermore, methods for increasing the therapy efficiency by using radio sensitizers and the state-of-the-art of defining precise patient treatment plans, identifying the clinical benefits of this type of therapy, were also addressed.

We would like to thank all participants for the lively exchange of ideas and results, thus making this conference a very fruitful event. Furthermore, we appreciate the financial support of the sponsors of this conference, in particular of the COST Action MP1002 financed by ESF. We would also like to express our thanks to all authors of these proceedings, as well as to the reviewers for their time, efforts and recommendations made during the preparation of this volume. Finally, many thanks to U G Huber for a careful proof-read of this manuscript.

We look forward to the 2nd Nano-IBCT Conference, which will be held in spring 2013.

Caen, 15 March 2012

Bernd A Huber, Christiane Malot, Alicja Domaracka and Andrey V Solov'yov
The Editors

Committees

The Conference has been organised in the framework of the COST Action MP1002 (Nano-IBCT: Nanoscale Insights into Ion Beam Cancer Therapy). Details can be found on the following website: http://fias.uni_frankfurt.de/nano-ibct

Co-chairs

Y Gauduel (ENSTA, Palaiseau, France)
B A Huber (CIMAP, Caen, France)
A V Solov'yov (FIAS, Frankfurt, Germany)

Local Organising Committee
(CIMAP, Caen, France)

L Adoui
A Domaracka
B A Huber
E Lattouf
S Maclot
C Malot
A Méry
J-C Poully
J Rangama
P Rousseau

Journal of Physics: Conference Series **373** (2012) 011001 doi:10.1088/1742-6596/373/1/011001

Sponsors

COST Action MP1002 NANO-IBCT

Conseil Régionale de Basse Normandie

Centre National de la Recherche Scientifique

Ecole Nationale d'Ingénieurs de Caen
& Centre de Recherche

Alliance Nationale pour les Sciences
de la Vie et de la Santé

Université de Caen Basse-Normandie

Virtual Atomic and Molecular Data Centre

Médicine & Lumière Synchrotron

Ecole Doctorale de l'Université
de Caen (SIMEM)

Pôle Normand des Sciences Nucléaires
& de leurs Applications

Pantechnik SA

RoentDek Handels GmbH

Ville de Caen

Centre de Recherche sur les Ions, les Matériaux
et la Photonique

Journal of Physics: Conference Series **373** (2012) 011001 doi:10.1088/1742-6596/373/1/011001

NANO-IBCT 2011, Caen
List of participants

N°	Family name	First name	Institution
1	ABRIL	Isabel	Universitat Alacant, Spain
2	ADOUI	Lamri	CIMAP-Caen, France
3	ARNDT	Alexander	Physikalisch Technische Bundesanstalt Braunschweig, Germany
4	AVALDI	Lorenzo	CNR-IMIP, Monterotondo Scalo, Italy
5	BACCARELLI	Isabella	CASPUR, Rome, Italy
6	BACCHUS	Marie-Christine	LASIM Lyon, France
7	BAEK	Woon Yong	Physikalisch Technische Bundesanstalt Braunschweig, Germany
8	BALANZAT	Emmanuel	CIMAP-Caen, France
9	BALD	Ilko	INC, Aarhus University, Aarhus, Denmark
10	Ban d Etat	Brigitte	CIMAP-Caen, France
11	BARI	Sadia	Max Planck Advanced Study Group, Hamburg, Germany
12	BECK	Peter	AIT Austrian Institute of Technology GmbH, Vienna, Austria
13	BEUVE	Michaël	Institut de Physique Nucléaire de Lyon, Villeurbanne, France
14	BOLOGNESI	Paola	CNR-IMIP, Monterotondo Scalo, Italy
15	BOUFFARD	Serge	CIMAP-Caen, France
16	BOULANOUAR	Omar	Université de Paris-Sud, Orsay, France
17	BUG	Marion	Physikalisch Technische Bundesanstalt Braunschweig, Germany
18	CAFARELLI	Pierre	IRSAMC LCAR, Université Paul Sabatier, Toulouse, France
19	CHAMPION	Christophe	Université Paul Verlaine-Metz, Metz, France
20	CASSIMI	Amine	CIMAP-Caen, France
21	CAUET	Emilie	Université Libre de Bruxelles, Bruxelles, Belgium
22	CHESNEL	Jean-Yves	CIMAP-Caen, France
23	CHIRIOTTO ALVAREZ	Sabine	SCK·CEN, Mol, Belgium
24	COLAUTTI	Paolo	INFN, Legnaro, Italy
25	COLLET	Solene	GIP Cycéron,Caen, France
26	COLLOC'H	Nathalie	GIP Cycéron,Caen, France
27	CONSTANS	Jean-Marc	Unité IRM CHU de Caen, France
28	D'AGOSTINO	Emiliano	Belgian Nuclear Research Center, Mol, Belgium
29	DENIFL	Stephan	University of Innsbruck, Austria
30	DE VERA	Pablo	Universitat d'Alacant, Alicant, Spain
31	DOMARACKA	Alicja	CIMAP-Caen, France
32	DOSANJH	Manjit	CERN, Genéve, Switzerland
33	DOS SANTOS	Morgane	IRSN, Fontenay-aux-Roses, France
34	DURANTE	Marco	GSI, Helmhotzzentrum, Darmstadt, Germany
35	DURANTEL	Florent	CIMAP-Caen, France
36	EDEN	Samuel	The Open University, Milton Keynes, United Kingdom
37	FALK	Martin	Institut of Biophysics of ASC, Brno, Czech Republic
38	FEKETEOVA	Linda	University of Melbourne, Melbourne, Australia
39	FRANCIS	Ziad	Saint Joseph University, Beirut, Lebanon
40	FRIEDLAND	Werner	Helmholtz Zentrum München, Germany
41	FUSS	Martina	Instituto de Física Fundmental (CSIC), Madrid, Spain
42	GARCIA	Gustavo	IFF-CSIC, Madrid, Spain
43	GARCIA-MOLINA	Rafael	Universidad de Murcia, Murcia, Spain
44	GAUDUEL	Yann	LOA, Ecole Polytechnique ENSTA, Palaiseau, France
45	GIGLIO	Eric	CIMAP-Caen, France
46	GONZALES-MANGANA	Olmo	KVI, Groningen, The Netherlands
47	GRYGIEL	Clara	CIMAP-Caen, France
48	HABERER	Thomas	Heidelberg Ion Therapy Center-University Hospital, Germany
49	HABRAND	Jean-Louis	Centre François Baclesse, Caen, France
50	HUBER	Bernd	CIMAP-Caen, France
51	HUELS	Michael	Sherbrooke University, Canada
52	ILLENBERGER	Eugen	Freie Universit¨t Berlin, Berlin, Germany
53	INGOLFSSON	Oddur	University of Iceland, Reykjavík, Iceland
54	KELLER	Adrian	Interdisciplinary Nanoscience Center, Aarhus, Denmark
55	KIRM	Marco	University of Tartu, Estonia
56	KOLBUN	Natalia	Physikalisch Technische Bundesanstalt Braunschweig, Germany

N°	Family name	First name	Institution
57	KOPYRA	Janina	Siedlce University, Siedlce, Poland
58	KOROL	Andrey	FIAS, Frankfurt, Germany
59	KRÄMER	Michael	GSI, Darmstadt, Germany
60	KUMAR	S.V.K.	Tata Institute of Fundamental Research, Mumbai, India
61	LACOMBE	Sandrine	Université de Paris-Sud, Orsay, France
62	LASTER	Brenda	Ben Gurion University - Beer Sheva, Israel
63	LATTOUF	Elie	CIMAP-Caen, France
64	MACLOT	Sylvain	CIMAP-Caen, France
65	MAECKEL	Volkhard	RIKEN, Wako, Japan
66	MAGGIORELLA	Laurence	NANOBIOTIX Paris, France
67	MAISONNY	Rémi	CIMAP-Caen, France
68	MALOT	Christiane	CIMAP-Caen, France
69	MANIL	Bruno	LPL - Université Paris 13 , France
70	MASON	Nigel	The Open University, Milton Keynes, United Kingdom
71	MATTHIEU	Germain	NANOBIOTIX Paris, France
72	MENDEZ	Louis	Universidad Autónoma de Madrid, Spain
73	MERY	Alain	CIMAP-Caen, France
74	METREVELI	Nunu	Ilia State University, Tibilisi, Georgia
75	MILOSAVLJEVIC	Alexandar	Institute of Physics Belgrade, Serbia
76	MOREELS	Marjan	SCK-CEN, Belgian Nuclear Research Center, Belgium
77	MORETTO-CAPELLE	Patrick	Université Paul Sabatier Toulouse, France
78	NAMISHEISHVILI	Louisa	Ilia State University, Tibilisi, Georgia
79	PAPP	Peter	Comenius University Bratislava, Slovakia
81	POTTIER	Agnès	NANOBIOTIX Paris, France
82	POULLY	Jean-Christophe	CIMAP-Caen, France
83	PRISE	Kevin	Queen's University Belfast, United Kingdom
84	RABADAN	Ismanuel	Universidad Autónoma de Madrid, Spain
85	RABUS	Hans	Physikalisch Technische Bundesanstalt Braunschweig, Germany
86	RADULESCU	Daniela	FIAS, Frankfurt, Germany
87	RANGAMA	Jimmy	CIMAP-Caen, France
88	ROLLET	Sofia	AIT Austrian Institute of Technology GmbH, Vienna, Austria
89	ROTHARD	Hermann	CIMAP-Caen, France
90	ROUSSEAU	Patrick	CIMAP-Caen, France
91	SCHETTINO	Guiseppe	Queen's University Belfast, United Kingdom
92	SCHLATHÖLTER	Thomas	KVI, Groningen, The Netherlands
93	SCIFONI	Emanuele	GSI, Helmoltzzenmtrum fuer Schwerionenforschung, Germany
94	SMIALEK-TELEGA	Malgorzata	Gdansk University of Technology, Gdansk, Poland
95	SMYTH	Maeve	Queen's University Belfast, United Kingdom
96	SOLOV'YOV	Andrey	FIAS, Frankfurt, Germany
97	STYPCZYNSKA	Agnieszka	The Open University, Milton Keynes, UK
98	SUETENS	Annelies	SCK-CEN, Belgian Nuclear Research Center, Belgium
99	SURAUD	Eric	IRSAMC Toulouse, France
100	SURDUTOVICH	Eugene	Oakland University-Rochester, USA
101	TANIS	John	Western Michigan University, Kalamazoo, USA
102	TESTA	Etienne	Institut de Physique Nucléaire de Lyon, France
103	TRIBEDI	Lokesh	Tata Institute of Fundamental Research, Mumbai, India
104	ULLRICH	Joachim	Max-Planck-Institut für Kernphysik, Heidelberg, Germany
105	VAN DER BURGT	Peter	National University of Ireland Maynooth, Ireland
106	VELTCHEVA	Mina	LOA, Ecole Polytechnique ENSTA, Palaiseau, France
107	VERKHOVTSEV	Alexey	FIAS, Frankfurt, Germany
108	VIBOK	Agnes	University of Debrecen, Hungary
109	VILLAGRASA	Carmen	IRSN-Fontenay aux Roses, France
110	WAELZLEIN	Cathrin	GSI Darmstadt, Germany
111	WHELAN	Caroline	COST Office, Brussels, Belgium
112	WHITLOW	Harry	University of Jyväskylä, Finland
113	YAKUBOVICH	Alexander	FIAS, Frankfurt, Germany
114	YAMAZAKI	Yasunori	RIKEN, Wako, Japan
115	ZYCHOR	Izabella	The Andrzej Soltan Institute for Nuclear Studies, Otwock, Poland

Peer review statement

All papers published in this volume of *Journal of Physics: Conference Series* have been peer reviewed through processes administered by the proceedings Editors. Reviews were conducted by expert referees to the professional and scientific standards expected of a proceedings journal published by IOP Publishing.

Nano-IBCT 2011 IOP Publishing
Journal of Physics: Conference Series **373** (2012) 012001 doi:10.1088/1742-6596/373/1/012001

Multiscale physics of ion-beam cancer therapy

E Surdutovich[1, 2] **and A V Solov'yov**[2,3]

[1] Department of Physics, Oakland University, Rochester, Michigan 48309, USA
[2] Frankfurt Institute for Advanced Studies, Ruth-Moufang-Str. 1, 60438 Frankfurt am Main, Germany
[3] On leave from A.F. Ioffe Physical Technical Institute, St. Petersburg, Russia

E-mail: `solovyov@fias.uni-frankfurt.de`

Abstract. We review a multiscale approach to the physics of ion-beam cancer therapy, an approach suggested in order to understand the interplay of a large number of phenomena involved in radiation damage scenario occurring on a range of temporal, spatial, and energy scales. We briefly overview its history and present the current stage of its development. The differences of the multiscale approach from other methods of understanding and assessment of radiation damage are discussed as well as its relationship to other branches of physics, chemistry and biology.

1. The roots of the multiscale approach

The scientific interest in obtaining a deeper understanding of radiation damage is motivated by the development of ion-beam cancer therapy and other applications of ions interacting with biological targets. A number of important scientific questions, especially related to DNA damage assessment on the molecular level, have not yet been resolved. Therefore, recently this field has attracted much attention from the scientific community [1–5]. There are a series of conferences devoted to these subjects, such as RADAM [1] and later Nano-IBCT [6]. The latter became possible due to the support of the European framework for Cooperation in Science and Technology (COST). The COST Action, "Nano-scale insights into ion beam cancer therapy (Nano-IBCT)" [7] was approved in 2010. Among these studies is the multiscale approach to the assessment of radiation damage induced by irradiation with ions. It is aimed at a phenomenon-based quantitative understanding of the scenario from the incidence of an energetic ion on tissue to the cell death. This method combines many spatial, temporal, and energy scales, and is therefore a truly multiscale approach.

1.1. Different approaches to radiation damage assessment

The understanding and assessment of radiation damage due to ionizing radiation are at the focus of radiation biophysics, which has a wide scope of important applications from radiation protection to radiation therapy. The standard scope of radiation biophysics spans from the interaction of ionizing projectiles with matter, radiation chemistry that includes interactions of radiation and secondary particles with water and biomolecules, to the analysis of models of cell survival [8–10]. The details of biodamage, such as different types of DNA lesions, single and double strand breaks (DSB) in DNA along with different pathways of their repair, are thoroughly discussed. The rest is devoted to biological effects, such as the variety of responses to different quantities or qualities of radiation on different levels from the subcellular to the organismic.

The physical part of this analysis indeed starts with a projectile entering tissue. These projectiles – x-rays, electrons, protons or heavier ions, neutrons, or other more exotic particles – are quite energetic in the beginning and ionization of the medium is the main channel of their energy loss. Therefore, these types of radiation are referred to as ionizing; even though there are other important channels of energy loss, such as nuclear fragmentation. Then, of course, there is a physical question "how does the radiation damage happen?" X-rays or γ-particles lose energy exponentially with respect to the traveled distance and this results in certain spectra and loci of secondary particles such as electrons, free radicals, etc. Other (massive) projectiles lose their energy differently and their energy loss per increment of trajectory $|dE/dx|$, called the linear energy transfer (LET), features a Bragg peak near the end of their path. This peak is the main physical reason for using particles as projectiles for radiotherapy, because the energy deposition in many cases can be localized, thus focusing on a tumor and sparing healthy organs. The next question is inevitably about the relation between energy deposition and the biological damage. This is where physics, chemistry, and biology overlap and this issue is arguably one of the most important in radiobiology.

One way to address the problem of biological damage starts from the analysis of so-called survival curves. The survival curve is a relation between the logarithm of cell-survival fraction and the deposited dose. These curves can be obtained experimentally and they differ for different species, cells, radiation modality, etc. Nevertheless, the main feature of these curves is that they are approximately parabolic, i.e.,

$$\ln S/S_0 = \alpha D + \beta D^2 \ , \tag{1}$$

where S/S_0 is the normalized cell survival, D is the dose, and α and β are coefficients. Several features of these curves have been discussed, one of which is the ratio of α/β. If this ratio is large, the survival curve is "steep" and more like a straight line; if it is small then it is a "shouldered" curve. Naturally, the next question is how to explain this, and there has been a series of models suggested since 1955 [8]. It is important to know that when a cell or a subcellular target like a cell nucleus, membrane, or mitochondrion is irradiated, there is an initial damage, e.g., a strand breaking in a DNA molecule, which can be described by atomic physics and chemistry. Then, there is a subsequent biological effort in order to repair the initial damage in which biological and biochemical mechanisms are invoked. The resultant damage revealed in survival curves is the result of these two phases. The models explaining survival curves relate the dose to the initial damage and its partial repair to obtain the parameters α and β in equation (1).

Typical statements resulting from applications of different models may sound like this: If there is no repair mechanism and every (single) hit of the target results in cell inactivation with a given probability, the ratio of α/β is very large, the first term on the right hand side of equation (1) is dominant, the survival curve is a straight line, i.e., inactivation is exponential with respect to the dose. Or if there is no repair and several hits are necessary to inactivate a cell, the curve is shouldered with zero derivative at $D = 0$. Or if there is repair, then the curve is always shouldered. For many practical purposes, an experimentally obtained curve, given by equation (1) is *sufficient* information for the evaluation of radiation damage, and it has been used for many years for treatment planning and optimization.

Notice, that in the above examples we never mentioned any atomic or molecular interactions; these and more information are hidden in the purely empirical coefficients α and β. If the dose distribution is uniform, it is possible to not open the "Pandora's box" of atomic/molecular physics, since it brings up too many difficult questions involving interactions with biomolecules that seem irrelevant as compared to the biological unknowns related to repair mechanisms. This was acceptable when only x-rays were used as projectiles. For x-rays, the dose distribution is uniform and treatment plans had to deliver certain doses to certain locations in order to achieve desired results.[1]

Particle projectiles change this picture. The dose distribution around each particle's path is highly nonuniform because of the number of secondary electrons released by the ionized molecules of the medium on every nm of the path. These electrons as well as holes and radicals comprise a complicated track structure with a radial distribution of the dose, while the space between different tracks is largely undisturbed.

A solution to this problem was suggested by the Katz approach in which the radial dose distribution is calculated and related to the inactivation of sub-cell-nucleus targets [11–14]. The quality factor of radiation was introduced in order to relate the survival curve parameters to a given radiation, differentiating between track types, inactivation modes, the structural complexity of targets, etc. The eventual goal of the Katz model was to calculate the relative biological effectiveness (RBE) [2,8,10], one of the key integral characteristics of the effect of ions compared to that of photons. This ratio compares the doses of different projectiles leading to the same biological effect. Nevertheless, the biological relation of the radial dose distribution with the cell survival probability was done based on the survival curves for x-rays, without analyzing particular physical processes, i.e., the empiric coefficients α and β remain central to this approach.

The Local Effect Model (LEM), developed at GSI[2], calculates the RBE assuming that the biological effect of radiation is entirely determined by the spatial distribution of the radial dose inside the cell nucleus. It relates the response of biological systems, following ion irradiation, to the corresponding response after x-ray irradiation [2,15]. Corrections for the quality of damage was included in a later version of the LEM [2]. This model operates on the schematic level using equation (1) with empirical coefficients α and β. The LEM solves technical problems related to the optimization of treatments, leaving no place for *ab initio* approaches and physical, chemical, or biological effects in general; even a consideration of DNA lesions such as DSBs is beyond the scope of LEM [1,2].

A defining feature of high-LET irradiation is not only the high concentration of dose *per se*, but the quality that this high concentration brings. In order to explain this, we have to explain some mechanisms of DNA damage. When tissue is irradiated with x-rays, they ionize and excite the medium and this results in a massive uniform production of hydroxyl radicals, which "attack" DNA molecules and induce damage in a series of chemical reactions. These processes were extensively studied in 1980s-1990s [16–18]. The concentration of events in such conditions is small and clustered damage is rare. The cell survival in this condition depends on the chemical environment (e.g., the presence of oxygen is deemed very important for radio resistivity), on how advanced the repair mechanisms in a given type of cells are, and on the stage in a cell cycle during which the primary damage occurs. A double strand breaking in DNA is the most severe lesion, but it can still be repaired [19–21]. The survival curves for many types of human cells are shouldered [10].

Particle projectiles, especially ions characterized with high LET, change this picture. A DNA segment sufficiently close to the ion's path can be hit a number of times. This causes complex damage, which is more lethal than a simple DSB [22]. This results in an obvious increase of the α/β ratio.[3] The LEM took this effect into account by introducing an extra empirical parameter [2,23], but this is just one example of the consequences of high-LET radiation! There are many other effects or possible environments, such as the application of sensitizers, which affect the number of secondary electrons, the application of an external magnetic field in the Bragg peak area, laser-driven beams, etc. It is hard to imagine how models, without theoretical

[1] We leave aside optimizations related to reducing dose deposition in healthy regions and treatment partitioning.
[2] Gesellschaft für Schwerionenforschung mbH, Darmstadt, Germany
[3] The mechanism of cell death is attributed to apoptosis [10].

Journal of Physics: Conference Series **373** (2012) 012001 doi:10.1088/1742-6596/373/1/012001

Table 1. Disciplines and scales of ion-beam cancer therapy

Phenomenon	Discipline	Space scale	Time scale
Beam generation	High energy physics	m–km	
Beam transport	Radiation physics	1–100 cm	10^{-7}s
Nuclear collisions and fragmentation	Nuclear physics	fm	10^{-22}s
Primary ionization, transport of secondaries	Atomic/molecular physics	0.1–10 nm	$10^{-17}-10^{-12}$ s
Branching of secondaries, radicals, excited species, chemical equilibrium	Chemistry	1–10 nm	$10^{-12}-10^{-5}$s
Local heating, heat transfer, stress	Thermo/hydro-dynamics	1–10 nm	$10^{-14}-10^{-9}$s
Dissociative electron attachment to molecules and other reactions	Quantum chemistry	Å	$10^{-15}-10^{-12}$s
Initial damage effects	Biochemistry	0.1–10 nm	10^{-5}s
Repairing mechanisms	Molecular biology	1–100 nm	s–min
Cellular network and interaction	Cell biology	μm	min
(Tumor) Cell death	Medicine	mm	min–years

base under the coefficients like α and β, can be sufficiently versatile to meet these challenges.

1.2. The alternative of the multiscale approach

The multiscale approach was formulated in references [4, 24], where a scenario leading to DNA damage was suggested. Then, it was elaborated upon as different aspects of the scenario were added in a series of work [5, 25–30]. Its name emphasizes the fact that important interactions involved in the scenario happen on a variety of temporal, spatial, and energy scales, as shown in table 1 [31].

Right from the beginning, the approach was formulated as phenomenon-based and was aimed at elucidating the physical, chemical, or biological effects that are important or dominating on each scale in time, space, and energy. The scheme of the multiscale approach is shown in figure 1 [4].

The formula, which defined the scenario, given in references [4, 24],

$$N_{DSB} = \Gamma_{DSB} \int dk \int d\mathbf{A} \cdot D\nabla P(k,r)\epsilon(k) \frac{dN}{d\zeta} d\zeta \ , \tag{2}$$

manifests the interplay of different scales. While the details of the derivation of this equation will be discussed below, notice that the equation predicts that the probability of a DSB (N_{DSB}) at a given segment of DNA is given by the probability of a DSB on impact by an electron, Γ_{DSB}, multiplied by the number of secondary electrons incident on this segment. That number is calculated as a result of secondary electron production ($\frac{dN}{d\zeta}d\zeta$) and diffusion ($D\nabla P(k,r)\epsilon(k)$) of these electrons to reach the target. Thus, a number of scales is included in this expression. The first calculation done using equation (2) analyzed the effect of orientation of a DNA twist with respect to the ion's path and it was concluded that the orientation might not be important for distances exceeding 2 nm, as shown in figure 2.

The multiscale approach raised questions about the nature of the effects that take place and lead to survival curves and the calculation of RBE and other macroscopic quantities,

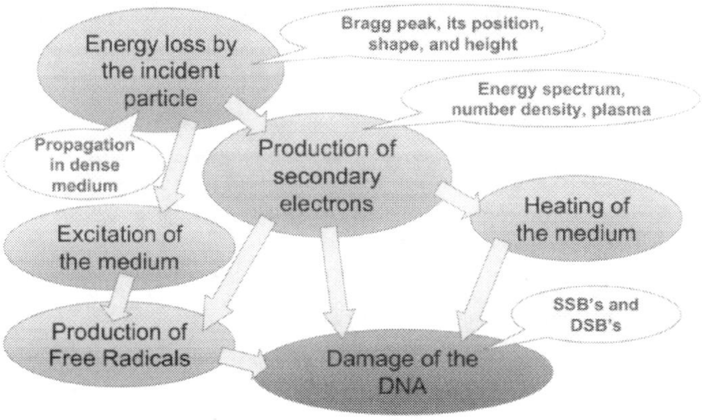

Figure 1. The scheme of the multiscale approach [4].

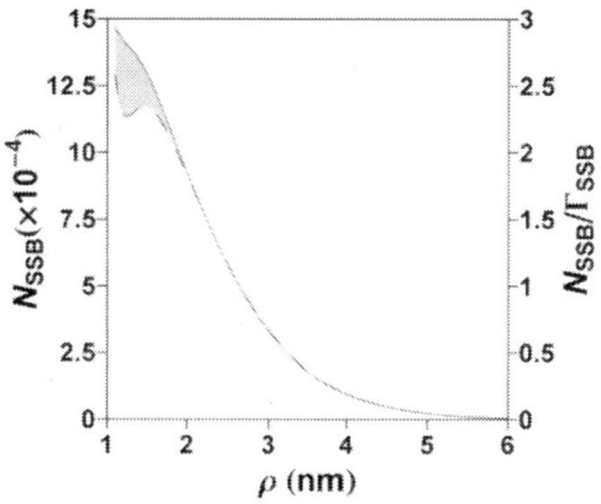

Figure 2. A comparison of dependencies of SSB yields on the distance between a DNA twist and the ion's path in the parallel (dashed line) and normal (solid line) cases [4].

e.g., by the end of the day it should answer the question of what is behind the coefficients in equation (1). The criticism that we often hear claims that our approach is not practical since the main questions have already been answered and the existing Monte Carlo (MC) simulations are capable of taking care of every single detail on every scale. Although we find simulations on different scales [32–38] to be very important in radiation damage assessment, we doubt both qualitative and quantitative comprehensiveness and exhaustiveness of the MC simulations. Two examples to this are the thermomechanical mechanisms discussed in references [28–30] and the double-electron-emission effects on DSBs discussed in reference [39]. Assurances about impracticality, infeasibility, and elusiveness have seldom stopped the scientific search for the truth.

2. A palette of physical effects on different scales

Here we review the achievements of the multiscale approach at different scales. The main issues addressed by the multiscale approach are ion stopping in the medium, the production and transport of secondary electrons produced as a result of ionization and excitation of the medium, the interaction of secondary particles with biological molecules, the most important being DNA, the analysis of induced damage, and the evaluation of the probabilities of subsequent cell survival or death. These effects are happening on time scales ranging from 10^{-21} to 10^{-5} s, i.e., from nuclear to biochemical times. The aim of the physical part of the analysis is the calculation of the spatial distribution of primary DNA damage, including the degree of complexity of this damage.

2.1. Ion stopping and the Bragg peak

The multiscale approach started with the analysis of ion propagation in a medium. Liquid water was used as the medium because human tissues on the average consist of 75% water. These work [25, 27] resulted in the description of the Bragg peak and the energy spectrum of secondary electrons. The Bragg peak in the stopping power of massive charged particles is obtained using a version of the Bethe - Bloch formula [40–42]. This formula provides the dependence of the stopping power on the energy of the ion and practically depends on the mean excitation energy. This energy for liquid water is chosen somewhere between 70 and 80 eV [36, 43]. Our

Journal of Physics: Conference Series **373** (2012) 012001 doi:10.1088/1742-6596/373/1/012001

approach to this problem was different. We chose to use the singly-differentiated (with respect to the secondary electron energy) ionization cross sections of water molecules in the medium as a physical input. These cross sections were taken from experiments [44] with parameters tuned for liquid water [27]. The parametrizations were also modified to take into account relativistic effects in the beginning of the ion's path. Even a slight change in cross sections in the entrance channel (plateau region of LET curve) may significantly affect the position of the Bragg peak. Energy loss due to excitation of water molecules has also been included in the calculations [25]. The effect of charge transfer strongly affects the height of the Bragg peak, since it is proportional to the square of the effective charge of the projectile and the latter decreases as the projectile slows down because of the pick off electrons. Barkas's parametrization [45] was used for the calculation of the effective charge. The next effect, included in reference [25], was the scattering of projectiles that naturally occurs as they propagate in the medium. This causes the spread in longitudinal velocities or the so-called energy straggling. This widens and substantially diminishes the Bragg peak for the beam of projectiles. Another effect on this scale, not yet included in the multiscale approach, is the nuclear fragmentation that happens quite often when projectiles collide with the nuclei of the medium. This effect is deemed to be significant, especially for the calculations of the tail in the stopping power curve beyond the Bragg peak [36]. As a consequence, we were able to describe the Bragg peak in good agreement with simulations and experiments [25–27] as shown in figure 3.

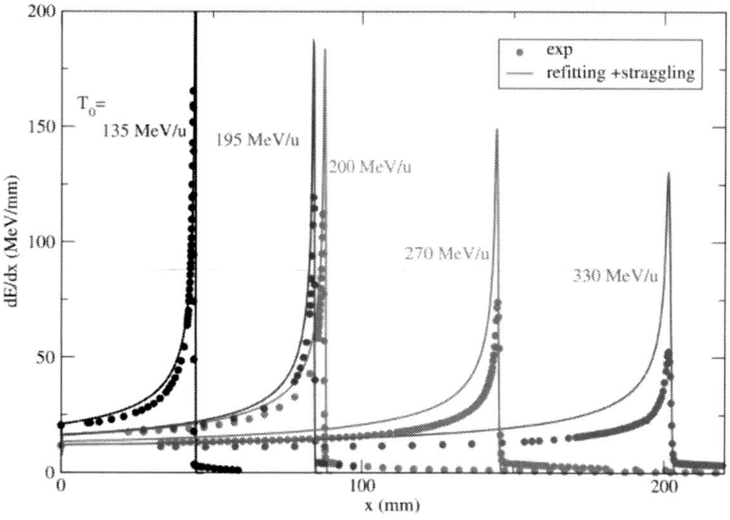

Figure 3. Linear energy deposition for carbon ions for different initial energies T_0: our model (all lines) compared to experiments from GSI [46] (all dots). Different labels and colors indicate curves at different initial ion-energies T_0 [27].

An important difference of our approach, relying on the singly-differentiated ionization cross section, allowed us to not only describe the Bragg peak, but also to use these cross sections to calculate the energy spectrum of secondary electrons, which turned out to be quite useful on the following scale of the electron transport and interactions of electrons with the medium.

2.2. Transport of secondary electrons and its relation to DNA damage

The next scale in energy and space is related to the transport of the secondary particles, which has been considered in references [4, 5, 39, 47]. These papers were built on the inference from the analysis of the energy distribution of secondary electrons that the average energy of these electrons is about 45-50 eV. At these and lower energies, the ionization cross sections are nearly isotropic, which allows one to employ a random walk approximation, i.e., assume that most of secondary electrons (excluding the more energetic δ-electrons) diffuse out from the ion's path. This diffusion was explored in a series of problems. First, we considered a random walk of electrons from the ion's path to a twist of DNA located at some distance from the path [4]. A single twist of a DNA molecule was chosen because it is an important unit for DSB analysis. This study predicted that the damage probability on impact does not strongly depend on the orientation of the DNA segment with respect to the ion's path. In that paper, we made an important and quite reasonable comparison of an estimate of the number of DSBs per μm of the ion's path with biophysical experiments [48, 49].

A later development of that work in reference [5] explored a random walk approximation in order to assess the complex DNA damage. Complex DNA damage, investigated by biologists in references [22, 50–52], is defined as a multiple number of primary DNA lesions happening on a length of two DNA twists. Such a correlation of lesions is called a cluster with a size equal to the number of singly-damaged sites. Different types of lesions may be qualified as effective, e.g., single strand breaks (SSBs), DSBs, base-damages, abasic sites, etc. The size of a cluster is related to the probability of its enzymatic repair. It is deemed that clusters of a sufficiently large size (larger than three or four lesions) are lethal for the cell. Thus, the problem of assessment of complex damage can, on the one hand, be tackled in the same fashion as the problem of DSB, described above; on the other hand, its solution is directly related to the assessment of cell survival as a result of irradiation with ions. The target for a complex damage site is a two-twist segment of a DNA molecule. The original idea for the calculation of complex damage was formulated in reference [47]. There, it was suggested that each lesion that counts in a cluster is due to an action of a single agent such as a secondary electron, a radical, etc. Then, if the average number of lesions per target N_i is calculated, the probability of occurrences of clusters of different size, $P_\nu(N)$, can be calculated using a Poisson distribution,

$$P_\nu = \exp\left(-N_i\right) \frac{N_i^\nu}{\nu!} \ , \tag{3}$$

where ν is the degree of complexity [5, 31]. Then, the problem of how to calculate N_i can be approached differently. The first method is to use the random walk approximation and calculate the number of particles (secondary electrons) incident on a given target as a consequence of irradiation or, much simpler, as a consequence of a single ion's traverse, as described below.

2.3. A random walk approach

The most general problem can be set up in the following way. We consider a three-dimensional axially symmetric random walk of secondary electrons from the axis, determined by the ion's path. The key quantity is the flux of secondary electrons through a patch $d\mathbf{A}$ located at a distance r from a segment $d\zeta$ of the ion's path from which the secondary electrons originate. It is given [5, 53, 54] by the following expression:

$$\begin{aligned} \frac{dN_A(\mathbf{r},t)}{dt} &= d\mathbf{A} \cdot D\nabla P(t,r) \frac{dN}{d\zeta} d\zeta \\ &= d\mathbf{A} \cdot D\mathbf{n_r} \frac{\partial P(t,r)}{\partial r} \frac{dN}{d\zeta} d\zeta \end{aligned} \tag{4}$$

where, $D = \bar{v}l/6$ is the diffusion coefficient, $\bar{v}$ is the average speed of the electron, l is the elastic mean free path of electrons in the medium, $\mathbf{n_r}$ is a unit vector in the radial direction (from the segment to the center of the patch $d\mathbf{A}$),

$$P(t,r) = \left(\frac{3}{2\pi\bar{v}tl}\right)^{3/2} \exp\left(-\frac{3r^2}{2\bar{v}tl}\right) \tag{5}$$

is the probability density to observe a randomly walking electron at a time t and a distance r from the electron's origin, and $\frac{dN}{d\zeta}$ is the number of secondary electrons emitted from a $d\zeta$-segment of the path.

It is beneficial to integrate equation (4) over time and then obtain the fluence through dA. In order to do this, we change variables from t to the number of steps done by secondary electrons k using $\bar{v}t = kl$. We rewrite equation (4), substituting (5), and switching from variable t to k as

$$dN_A(\mathbf{r}) = \int \frac{dN_A(\mathbf{r},t)}{dt} dt = d\mathbf{A} \cdot \mathbf{n_r} \frac{dN}{d\zeta} d\zeta \int_{r/l}^{\infty} dk \frac{r}{2k} \left(\frac{3}{2\pi k l^2} \right)^{3/2} \exp\left(-\frac{3r^2}{2kl^2} - \gamma k \right) . \quad (6)$$

An attenuation factor, $e^{-\gamma k}$, is introduced to account for electrons falling out from the random walk. The constant γ is equal to the ratio of the cross section of electron-inactivating processes to the total cross section. The integration over k is carried out from the minimal number of steps, necessary to reach distance r, to infinity.

The last integration is done over $d\mathbf{A}$ and it depends on the particular situation. As a result, the average number of incidences on the target N_A is calculated and this information may be quite useful [39]. N_i is calculated as a product of N_A and the probability of ionization or any other qualified damage on impact.

A number of problems can be solved using this approach. Two of them, the investigation of SSB and DSB yields and the assessment of complex damage, were considered in references [4,5]. An example of the dependence of cluster damage probabilities on the radial distance from the ion's path, calculated using this method, is shown in figure 4 [5]. Another important investigation is presented in reference [54]. The purpose of this study was to compare the random walk approach with MC simulations for nano-dosimetric applications. In this study, the biomolecules are not involved and N_i is the number of ionizations in the target. This number can also be measured or related to measurements. Such a study provides a way of validating our random walk approach and makes this method useful to the nano-dosimetry community [37,55].

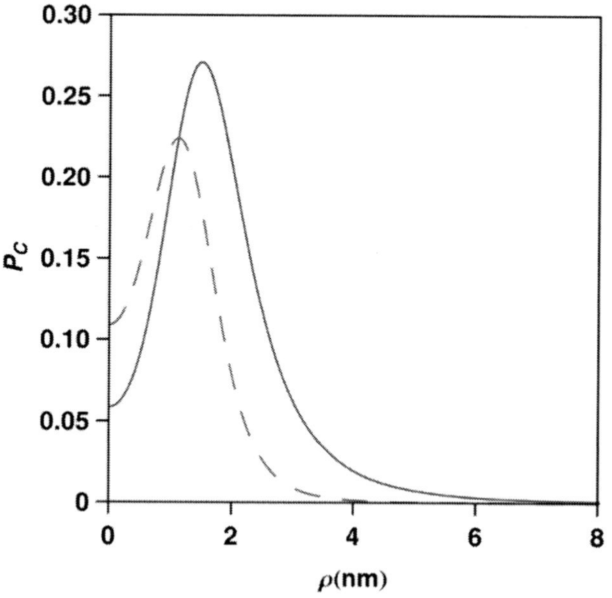

Figure 4. An example of radial distribution of clusters of two (solid line) and clusters of three lesions (dashed line) [5].

Still another application of the random walk can be found in the analysis of the relation of double ionization events to DSBs.

2.4. Double electron emission and DSBs

If the primary ionizing projectiles are ions, a substantial fraction of damage is done by secondary electrons, formed in the process of the ionization of the medium. A number of different pathways of damage due to these electrons were considered in references [3,5,19]. The direct measurements of DNA damage due to incident electrons were presented in, e.g., reference [56], which brought to light the possibility that low energy electrons were important agents of DNA damage. Since then, the mechanism of a SSB due to the action of a single electron, related to the formation of a transient negative ion (TNI) as a part of the process of dissociative electron attachment (DEA), has been widely discussed in the literature, e.g., references [3,57,58] with emphasis on low energy (under ionization threshold) of the incident electrons.

Unexpectedly high yields of DSBs compared with SSBs, in reported reference [56], lead to an hypothesis that DSBs can be caused by a single electron. This logic has been widely accepted and used in reference [4]. However, the mechanism of DSBs due to low-energy electrons is still quantitatively unclear despite qualitative arguments, suggesting that the breaks in the second strand are due to the action of debris generated by the first SSB [56]. In reference [4], the production of DSBs by two separate electrons was also considered, but that analysis was then shelved, since the number density of secondary electrons due to the primary ionization with an ion was not nearly enough for this effect to be considerable in comparison with DSBs due to a single electron action.

In reference [39], we argued that additional electrons emitted in the vicinity of a DNA molecule as a result of double-electron-emission events increase the above effect and thus constitute a mechanism for a DSB. We explored the probability of two electrons, produced in the vicinity of a DNA molecule, be incident on a single twist of this molecule. These additional electrons emerge as a result of double ionization events such as the Auger effect in single molecules, e.g., references [59–62] for water, or due to the effect of intermolecular coulombic decay (ICD), studied in references [63–67] for water clusters. The actual mechanism of double ionization is not important for this analysis. We used a generic term "double ionization events" to describe all relevant events leading to a production of additional electrons.

Auger electrons are produced as a result of non-radiative relaxation of holes produced by primary and secondary ionization. They may emerge consequent to the ionization of water or other molecules or clusters of the medium. If a secondary electron ionizes a molecule by kicking out one of its inner-shell electrons, a hole is formed on this molecule. If this hole then relaxes via a non-radiative channel (an Auger electron is emitted from the same molecule or an adjacent molecule in a cluster) the total number of electrons, emerging from a sub-nm locality, is equal to *three*, the ionizing electron, the released electron, and the Auger electron, as shown in reference 5. If a double-ionization event occurs on a DNA molecule, the probability of damage, such as a

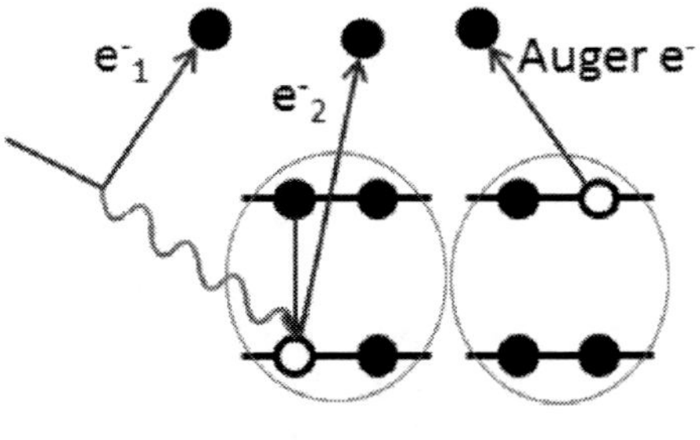

Figure 5. The scheme of ionization with the ICD; e_2 and the Auger electron emerge from different molecules. [39].

DSB, comprises the sequence of these processes with the corresponding dynamics of the DNA molecule and possible further interaction with these three electrons. If a water molecule (cluster) of the medium located in the vicinity of a DNA molecule is a host of a double-ionization event, then these three electrons can independently diffuse and stumble onto a DNA convolution and produce two SSBs, possibly yielding a DSB. We considered the latter scenario in more detail.

It is interesting to notice that the energies of each of the three electrons, emerging from a water molecule or cluster, are likely to be below 15 eV. An average ionizing electron has an energy around 45-50 eV, corresponding to the average energy of electrons produced by an ion [25]. After this ionization event, it loses about 30 eV [64].[4] Thus, the ionizing electron is left with less than 15 eV while the other two will have even smaller energies [65]. These low-energy electrons are likely to be engaged into the DEA channel leading to SSBs. How does this scenario make a difference in the DSBs production?

The number density of secondary electrons, produced on the ion's path, reduces rather steeply with the increasing distance from the path, so that the probability of two electrons incident on a single convolution of a DNA molecule, located a few nm from the path is very small. However, when ionization of a water molecule (cluster) producing extra electrons occurs at a distance from the path, three electrons emerging from the same spot substantially boost the local number density of electrons and, hence, the probability of a nearby DNA convolution to be hit with two electrons. A similar process may take place if the primary projectile is a photon. In that case, two (instead of three) electrons are produced in a locality and are capable of producing a DSB in a DNA molecule. If incident photons ionize the $1a_1$ state, then a cascade of Auger electrons may follow, rapidly increasing the number density of electrons.

In order to give a quantitative example of the effect caused by double-ionization events, we considered a double ionization event caused by a secondary electron, produced by an incident carbon ion. We considered a random walk of secondary electrons from the point of the event and calculated the probability that two of these electrons hit a given DNA convolution. The results of these calculations are presented in figure 6.

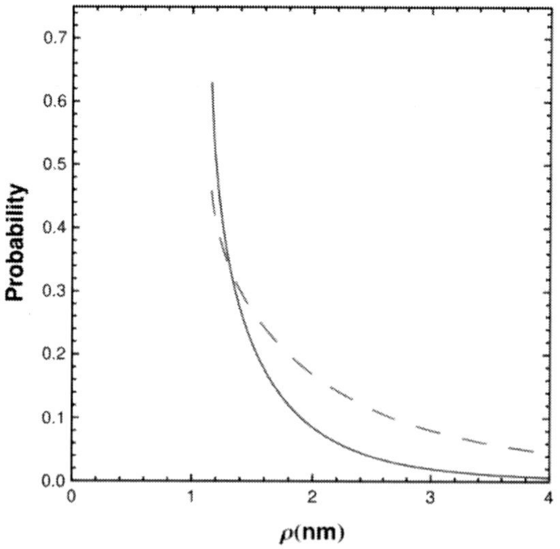

Figure 6. The probability for two electrons to pass through a single convolution of DNA (solid line) compared to that of one electron (dashed line) [39].

It is remarkable, that for $r < 2$nm the probability of the impact of two electrons is comparable to that of one. Of course, figure 6 includes neither the value of the probability of DEA on impact, nor the probability of a double electron emission event, but still, the substantial quantity of this fluence elevates the interest to study such events and their influence on DNA damage.

[4] The discussed double-ionization events do not concern electrons from the $1a_1$ MO of an H_2O molecule, because in order to kick these electrons out, the secondary electron has to have an energy above 540 eV. Such electrons are very rare.

The conclusion of that paper is that due to double-electron emission events, two or more electrons interact with a single DNA convolution with a sizable probability. Double strand breaks or other types of complex damage may result from these interactions. Since the Auger emission may result from the ionization of a water molecule by an incident photon, the above two-electron mechanism can also contribute to the DSB yield consequent to an irradiation of tissue with photons.

We considered the major concepts related to the transport of extra electrons and made first steps to including these electrons in the multiscale approach for the calculation of radiation damage by ions. The inclusion of Auger-like effects into the multiscale approach is an important step, since they play a significant role in radiation damage [68].

Future research in this route will be based on the calculations and measurements of the cross sections of double ionization of water molecules and clusters and the cross sections of DSBs due to two incident electrons. The calculation of the fluence of extra electrons, also presented in the paper, gives a framework for the calculation of the transport of free radicals formed due to an ion's traverse through a medium.

2.5. Radial dose based approach

There is an alternative to the calculation described above for the fluence using the random walk approach in determining the average number of lesions or ionizations per target. This approach is based on the calculation of the radial dose and integrating it over the target volume. This approach is somewhat reminiscent of the Katz approach, except the radial dose is calculated using the random walk approximation. The radial dose was first calculated using this method in reference [47] and then improved in reference [5]. The state of the art is presented in reference [69].

The calculations performed using the random walk approach are based on the integration of equation (6) over the area of a cylinder coaxial with the ion's path. The number of ionization events in a cylindrical slice between ρ and $\rho + d\rho$ is proportional to the number of secondary particles (of a given energy) incident on the inner cylindrical surface multiplied by the probability of ionization per electron (effective area over total area), which is equal to the number of water molecules inside the volume (number density times volume $n2\pi\rho L d\rho$) multiplied by the ionization cross section σ and divided by the total area of the cylinder $2\pi\rho L$,

$$d\mathcal{N} = N_A(\rho)\frac{n\sigma 2\pi\rho L d\rho}{2\pi\rho L} = N_A(\rho) n\sigma d\rho \ . \tag{7}$$

The energy, $d\mathcal{E}$, deposited in this slice is equal to the product of this number of events by the average energy per event $\bar{W}$:

$$d\mathcal{E} = \bar{W} N_A(\rho) n\sigma d\rho \ . \tag{8}$$

Finally, the radial dose, $d\mathcal{D}$, is the volume density of the deposited energy, i.e., equation (8) divided by the volume of the slice $2\pi\rho L d\rho$:

$$\mathcal{D} = \bar{W}\frac{N_A(\rho) n\sigma d\rho}{2\pi\rho L d\rho} = \bar{W}\frac{N_A(\rho)\sigma n}{2\pi\rho L} \ . \tag{9}$$

The radial dose, given by equation (9) can be compared to the experimental measurements and simulations. Then, the number N_i can be obtained by dividing equation (9) by $\bar{W}$.

3. Thermomechanical damage

In this section, we describe a study of the possibility of producing DNA damage as a result of a thermomechanical effect caused by ions stopping in a tissue. Thermomechanical effects in this case can be described as a dynamical change in temperature and pressure in the medium causing forces that may rupture bonds in DNA molecules [70]. The understanding of such a possibility evolved from the estimates of the temperature increase in the medium as a result of ion propagation [25], then from the analysis of thermal and pressure spikes in liquid water [28].

As we discussed above, an ion passing through the medium loses its energy. The energy loss is quantified by the LET and occurs mainly through ionization of the medium, and the energy is carried away by the secondary electrons produced in this process. Secondary electrons lose their acquired energy in a series of consecutive collisions with the molecules of the medium (predominantly water), which are likely to get excited in this process. Finally, these molecules relax and most of their energy is distributed between their translational degrees of freedom; the temperature inside a sub-nm cylinder surrounding the ion's path becomes very high. These predictions were made as a result of the calculations of heat transfer in the vicinity of ion paths in water using the inelastic thermal spike model [28].

A preliminary study of a biomolecule, a small protein ubiquitin, exposed to transient heating was carried out in reference [31]. That study showed that the temperature spike caused by a carbon ion "unglues" a protein molecule. A further dynamics study was not feasible until the pressure development scenario was understood in reference [29]. That work predicted that a sharp increase of temperature and pressure in the vicinity of the ion's path causes a shockwave, characterized by the rapid cylindrical expansion with a steep rise of pressure on the wavefront and its drop in the wake. The dependence of the pressure at the wavefront on its distance from the ion's path for a carbon ion is shown in figure 7 [29]. It was shown that forces, caused by

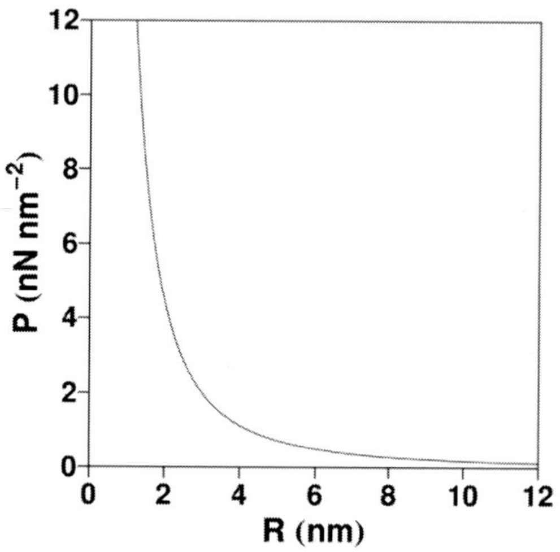

Figure 7. The dependence of the pressure on the shockwave front from its distance from the ion's path [29].

high-pressure gradients emerging as a result of local heating of the medium by passing ions, can be strong enough to break covalent bonds (more than 10 nN), but act only for a very short time and it remained unclear whether this was sufficient to break covalent bonds and cause severe damage to DNA molecules. This question was analyzed in references [30, 71] using a Molecular Dynamics (MD) simulation of a shockwave propagating through a nucleosome.

The results of the MD simulations demonstrate that the three dimensional structure of a nucleosome experiences noticeable obliteration as shown in figure 8 [30]. The propagating shockwave distorts the stacking interactions and the hydrogen bonds stabilizing the nucleotides of the DNA and leads to the partial destruction of the DNA secondary structure.

The analysis of stresses on the covalent bonds of the DNA backbone chain during the shock

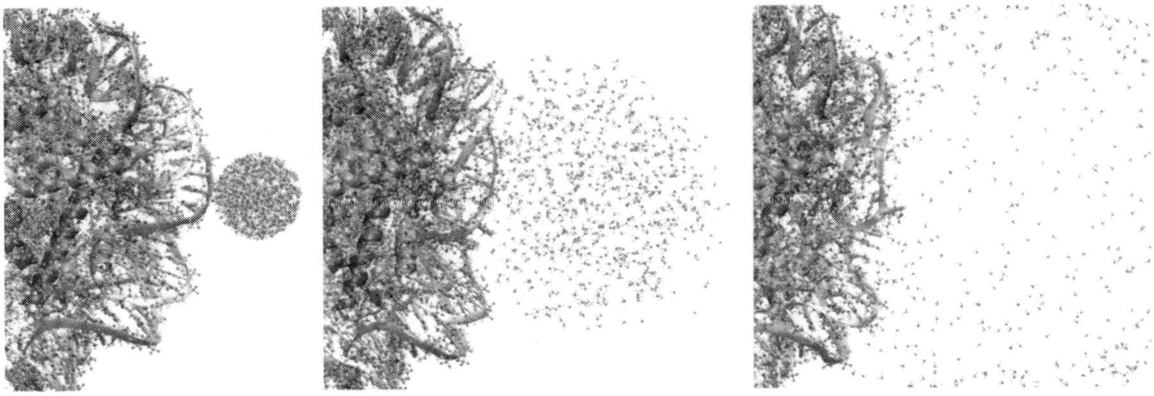

Figure 8. Structure of nucleosome after 0, 2, and 10 ps after carbon ion passage (perpendicular to the page). "Hot" water molecules being initially located within 1 nm of the ion's track are shown in the figure as an "exploding" cylinder to the right from the nucleosome. A substantial distortion of the nucleosome's three-dimensional structure caused by the shockwave is clearly seen [30].

wave propagation shows that the probability to induce a thermomechanical breakage of the DNA backbone chain in a single event of a carbon ion's close passage (1 nm from the surface of nucleosome) is equal to 0.5% [30]. This number significantly increases as the distance between the trajectory and the nucleosome decreases and/or with the increase of the effective charge of the ion [71, 72]. Thus, for ions such as iron, the thermomechanical mechanism of DNA damage may become dominant.

4. Conclusion

We have reviewed the multiscale approach to radiation damage by ions. We showed that the main difference from other approaches is the in-depth focusing on physical effects that we call a phenomenon-based approach. The state of the art of this approach is discussed with demonstration of some techniques of calculations. This is the first comprehensive review of our approach and we hope that its development will continue along the lines indicated in this review.

The main advantages of the multiscale approach follow from its architecture and are its fundamentality and versatility. The approach evaluates the relative contributions and significance of a variety of phenomena; it elucidates a complex multiscale scenario in sufficient detail and has a solid predictive power. It is structurally simple and inclusive, and allows for modifications and extensions by including new effects on different scales and improvements on the way.

Examples of innovative developments that we have discussed are the relation of double-ionization events with the nature of DSBs and thermomechanical mechanisms of biodamage. However, there are more to come. There are effects related to the action of sensitizers, to hypoxic/aerobic environments, and other conditions which can be treated as different phenomena and we believe that such treatments can be beneficial. A series of new opportunities will be related to the imminent proton beams produced by high-power lasers [73]. The calculation of RBE in such environments will require a more physical input.

Certainly, our goal and hope in developing the multiscale approach is to make a worthwhile tool for the assessment of radiation damage on the molecular level. While its practicality is still in question, its fundamental basis and depth related to atomic/molecular physics is becoming more and more evident.

Acknowledgments
We are grateful to J. Scott Payson and Bradley Roth who critically read the manuscript and the support of COST Action MP1002 "Nano-scale insights into ion beam cancer therapy."

References
[1] Baccarelli I, Gianturco F, Scifoni E, Solov'yov A and Surdutovich E 2010 *Eur. Phys. J.* D **60** 1
[2] Schardt D, Elsässer T and Schulz-Ertner D 2010 *Rev. Mod. Phys.* **82** 383–425
[3] Kumar A and Sevilla M 2010 *Chem. Rev.* **110** 7002–7023
[4] Solov'yov A, Surdutovich E, Scifoni E, Mishustin I and Greiner W 2009 *Phys. Rev.* E **79** 011909
[5] Surdutovich E, Gallagher D C and Solov'yov A V 2011 *Phys. Rev.* E **84** 051918
[6] URL `http://nano-ibct.sciencesconf.org/`
[7] URL `http://fias.uni-frankfurt.de/nano-ibct/`
[8] Alpen E L 1998 *Radiation Biophysics* (San Diego, London, Boston, New York, Sydney, Tokyo, Toronto: Academic Press)
[9] Lehnert S 2008 *Biomolecular Action of Ionizing Radiation* (New York,London: Taylor & Francis)
[10] Hall E J and Giaccia A J 2012 *Radiobiology for Radiologist* (Philadelphia, Baltimore, New York, London: Lippincott Williams & Wilkins)
[11] Butts J J and Katz R 1967 *Radiat. Res.* **30** 855–871
[12] Katz R, Ackerson B, Homayoonfar M and Sharma S C 1971 *Radiat. Res.* **47** 402–405
[13] Cucinotta F, Nikjoo H and Goodhead D 1999 *Radiat. Environ. Biophys.* **38** 81–92
[14] Korcyl M and Waligórski M 2009 *Int. J. Radiat. Biol.* **85** 1101–1113
[15] Scholz M and Kraft G 1996 *Adv. Space Res.* **18** 5–14
[16] von Sonntag C 1987 *The chemical basis of radiation biology* (London: Taylor & Francis)
[17] Sevilla M, Becker D and Razskazovskii Y 1997 *Nucleonika* **42** 283–292
[18] Becker D, Adhikary A and Sevilla M 2010 *Charged Particle and Photon Interactions with Matter Recent Advances, Applications, and Interfaces* (Boca Raton: CRC Press, Taylor & Francis)
[19] Kiefer J 1990 *Biological Radiation Effects* (Berlin, Heidelberg, New York: Springer-Verlag)
[20] Kanaar R, Hoeijmakers J and van Gent D 1998 *Trends Cell Biol.* **8** 483
[21] Becker D and Sevilla M 1993 *Advances in Radiation Biology* vol **17** ed Lett J (Acad. Press) pp 121–180
[22] Sage E and Harrison L 2011 *Mutat. Res.* **711** 123–133
[23] Elsaesser T and Scholz M 2007 *Radiat. Res.* **167** 319–329
[24] Surdutovich E and Solov'yov A 2009 *Europhys. News* **40/2** 21
[25] Surdutovich E, Obolensky O, Scifoni E, Pshenichnov I, Mishustin I, Solov'yov A and Greiner W 2009 *Eur. Phys. J.* D **51** 63–71
[26] Surdutovich E, Scifoni E, and Solov'yov A 2010 *Mutat. Res.* **704** 206–212
[27] Scifoni E, Surdutovich E and Solovyov A 2010 *Phys Rev.* E **81** 021903
[28] Toulemonde M, Surdutovich E and Solov'yov A 2009 *Phys. Rev.* E **80** 031913
[29] Surdutovich E and Solov'yov A 2010 *Phys. Rev.* E **82** 051915
[30] Yakubovich A V, Surdutovich E and Solov'yov A V 2012 *Nucl. Instr. Meth.* B **279** 135–139
[31] Surdutovich E, Yakubovich A and Solov'yov A 2010 *Eur. Phys. J.* D **60** 101
[32] Nikjoo H, Uehara S, Emfietzoglou D and Cucinotta F A 2006 *Radiat. Meas.* **41** 1052
[33] Friedland W, Jacob P, Bernhardt P, Paretzke H and Dingfelder M 2003 *Radiat. Res.* **159** 401–410
[34] Nikjoo H, Bolton C, Watanabe R, Terrisol M, PO'Neill and Goodhead D 2002 *Radiat. Prot. Dosim.* **99** 77–80
[35] Dingfelder M, Ritchie R, Turner J, Friedland W, Paretzke H and Hamm R 2008 *Radiat. Res.* **169** 584–594
[36] Pshenichnov I, Mishustin I and Greiner W 2008 *Nucl. Inst. Meth.* B **266** 1094–1098
[37] Bug M, Gargioni E, Guatelli S, Incerti S, Rabus H, Schulte R and Rosenfeld A 2010 *Eur. Phys. J.* D **60** 85–92
[38] Hsiao Y and Stewart R 2008 *Phys. Med. Biol.* **53** 233–244
[39] Surdutovich E and Solov'yov A V 2012 *Eur. Phys. J.* D submitted
[40] Bethe H 1930 *Ann. Phys.* **397** 325–400
[41] Bloch F 1933 *Z. Phys. A: Hadrons Nucl.* **81** 363–376
[42] Bloch F 1933 *Ann. Phys.* **408** 285–320
[43] Abril I, Garcia-Molina R, Denton C, Kyriakou I and Emfietzoglou D 2011 *Radiat. Res.* **175** 247–255
[44] Rudd M E, Kim Y K, Madison D H and Gay T 1992 *Rev. Mod. Phys.* **64** 441
[45] Barkas W H 1963 *Nuclear Research Emulsions I. Techniques and Theory* vol 1 (New York, London: Academic Press)
[46] Haettner E, Iwase H and Schardt D 2006 *Rad. Protec. Dosim.* **122** 485
[47] Scifoni E, Surdutovich E and Solov'yov A 2010 *Eur. Phys. J.* D **60** 115

[48] Jakob B, Scholz M and Taucher-Scholz G 2003 *Radiat. Res.* **159** 676–684
[49] Tobias F, Durante M, Taucher-Scholz G and Jakob B 2010 *Mutat. Res.* **704** 54–60
[50] Ward J 1988 *Prog. Nucleic Acid. Res. Mol. Biol.* **35** 95–125
[51] Ward J 1995 *Radiat. Res.* **142** 362–368
[52] Malyarchuk S, Castore R and Harrison L 2009 *DNA Repair* **8** 1343–1354
[53] Chandrasekhar S 1943 *Rev. Mod. Phys.* **15** 1
[54] Bug M, Surdutovich E, Rabus H, Rosenfeld A B and Solov'yov A V 2012 *Eur. Phys. J.* D submitted
[55] Grosswendt B 2005 *Radiat. Prot. Dosim.* **115** 1–9
[56] Huels M A, Boudaïffa B, Cloutier P, Hunting D and Sanche L 2003 *JACS* **125** 4467
[57] Gianturco F A, Sebastianelli F, Lucchese R R, Baccarelli I and Sanna N 2008 *J. Chem. Phys.* **128** 174302
[58] Park Y, Li Z, Cloutier P, Sanche L and Wagner J 2011 *Radiat. Res.* **175** 240–246
[59] Moddeman W, Carlson T, Krause M and Pullen B 1971 *J. Chem. Phys.* **55** 2317
[60] Siegbahn H, Asplund L and Kelfve P 1975 *Chem. Phys. Lett.* **35** 330
[61] Hjelte I, Piancastelli M, Fink R, Björneholm O, Bässler M, Feifel R, Giertz A, Wang H, Wiesner K, Ausmees A, Miron C, Sorensen S and Svensson S 2001 *Chem. Phys. Lett.* **334** 151–158
[62] Mitani M, Takahashi O, Saito K and Iwata S 2003 *J. Elec. Spect. Relat. Phen.* **128** 103–117
[63] Cederbaum L S, Zobeley J and Tarantelli F 1997 *Phys. Rev. Lett.* **79** 4778
[64] Müller I B and Cederbaum L S 2006 *J. Chem. Phys.* **125** 204305
[65] Stoychev S D, Kuleff A I and Cederbaum L S 2011 *J. Am. Chem. Soc.* **133** 6817–6824
[66] Jahnke T, Sann H, Havermeier T *et al.* 2010 *Nat. Phys.* **6** 139–142
[67] Mucke M, Braune M, Barth S, Förstel M, Lischke T, Ulrich V, Arion T, Becker U, Bradshaw A and Hergenhahn U 2010 *Nat. Phys.* **6** 143–146
[68] McMahon S, Hyland W, Muir M, Coulter J, Jain S, Butterworth K, Schettino G, Dickson G, Hounsell A, O'Sullivan J, Prise K, Hirst D and Currell F 2011 *Sci. Rep.* **1** 18
[69] Surdutovich E and Solov'yov A V 2012 *Eur. Phys. J.* D submitted
[70] Beyer M and Clausen-Schaumann H 2005 *Chem. Rev.* **105** 2921
[71] Yakubovich A V, Surdutovich E and Solov'yov A V 2012 in preparation
[72] Yakubovich A V, Surdutovich E and Solov'yov A V 2012 *These Proceedings* submitted
[73] Bulanov S S, Brantov A, Bychenkov V Y, Chvykov V, Kalinchenko G, Matsuoka T, Rousseau P, Reed S, Yanovsky V, Krushelnick K, Litzenberg D W and Maksimchuk A 2008 *Med. Phys.* **35** 1770

Action of secondary ions on biomolecules: anisotropy and radio-sensitization properties

M C Bacchus-Montabonel

Laboratoire de Spectrométrie Ionique et Moléculaire, Université de Lyon (Lyon I), CNRS-UMR5579, 43 Bd. du 11 Novembre 1918, F-69622 Villeurbanne Cedex, France

e-mail: bacchus@lasim.univ-lyon1.fr

Abstract. The charge transfer process of carbon ions on biomolecular targets is analyzed with regard to the action of radiation on the biological medium. The theoretical treatment is performed in a wide collision energy range by means of *ab-initio* quantum chemistry molecular methods. The process has been investigated for a series of target molecules, thymine, uracil and 5-halouracil corresponding to a similar skeleton with different substituent. The charge transfer appears markedly anisotropic in the whole energy domain, and interesting specific features may be pointed out at low energies. In addition, such a study may provide information on the radio-sensitivity of the different bases with regard to ion-induced radiation damage.

1. Introduction

The interaction of ionizing radiation with biological tissues can induce severe damage to DNA [1] as single- and double-strand breaks which can create important biological lesions and may lead to cancer. Apart from the radiation, secondary particles, electrons, or ions generated along the track after interaction of the ionizing radiation with the biological medium, have been shown to induce significant damage [2]. Numerous studies focused on the behaviour of the DNA building blocks under irradiation with slow-electrons, photons or ions with the aim to analyze the mechanisms involved at the molecular level. Low-energy electrons have been extensively investigated and they are shown to induce important damage even at very low kinetic energies, down to 3eV [3-6]. However, experimental and theoretical studies have also been performed on interactions of ions with the biological medium, first of all in the keV energy range corresponding to the region of the Bragg peak where the damage is maximum [7-16] and makes heavy-ion therapy such a promising technique in cancer treatment [17]. More recently, studies have been extended to lower collision energies [18-22] where specific physico-chemical interactions with the biological medium can occur. Collisions of ions with biomolecular targets may induce different processes: excitation and fragmentation of the biomolecule, ionization of the gaseous target, and also charge transfer from the multiply charged ion towards the biomolecule [23]. Experimentally, excitation and fragmentation cross sections are determined from mass spectra [7-9]. But theoretically, the charge transfer process may be investigated in the framework of the molecular representation of the collisions [24]. Fragmentation and charge transfer processes have been shown to be complementary processes and a strong influence of the charge of the projectile has been pointed out [11]. We have thus investigated the collision of the same ion on a series of biomolecular targets of similar structure, the DNA base thymine, the RNA base uracil and the corresponding 5-halouracil molecules corresponding to the same skeleton, but with a different substituent on the carbon C5 (see figure 1b). This extended study allows the analysis of the charge transfer mechanism with regard to both steric and electronic effects as groups of different size and electro-negativity are involved. We have chosen for this study the C^{4+} projectile ion which was shown to induce efficient charge transfer, weakly dependent on the collision energy [7,11], and which presents similar features as the C^{6+} ion used in hadrontherapy treatments [25]. The calculation has been performed in a wide collision energy range, from keV incident ions where incident damage is supposed to be maximum, to lower collision energies in order to exhibit a possible specific behaviour which could be of importance in ion-induced radiation damage. We focused our attention on the anisotropy of the process with regard to the substituent, as well as on the radio-sensitivity properties of halouracil molecules widely used in radiation therapy [26]. *Ab-initio* quantum chemistry methods have been used for the determination of the potential energies and non-adiabatic coupling matrix elements followed by a semi-classical collision treatment.

2. Theoretical approach

The charge transfer is described as the evolution of the quasi-molecular ion-target system. For such a polyatomic collision system, a simple model may be proposed by means of the one-dimensional reaction coordinate approximation [27-29], the polyatomic C^{4+} - biomolecule complex being considered as a pseudo-diatomic molecule whose evolution is driven by the reaction coordinate corresponding to the distance between the centre-of-mass of the target and the colliding carbon ion. Of course, such an approach does not consider the degrees of freedom of the complex and the internal motions of the molecule but it appears reasonable for very fast collision processes where nuclear vibration and rotation periods are assumed to be much longer than the collision time.

The geometry of the collision system is presented in figure 1a,b. The different molecular states involved in the charge transfer process are calculated along the reaction coordinate R for different approaches θ from the perpendicular (θ=90°) to a planar or near-planar geometry in order to take into account the anisotropy of the process. The potentials have been determined for a large number of R distances, from 0.5Å to 9Å, for a number of specific values of the angle θ. The angle φ has been kept fixed at φ=60° which corresponds to a direction opposite to the X substituent. The geometry of the ground state of the different biomolecular targets have been optimized and kept frozen during the collision process.

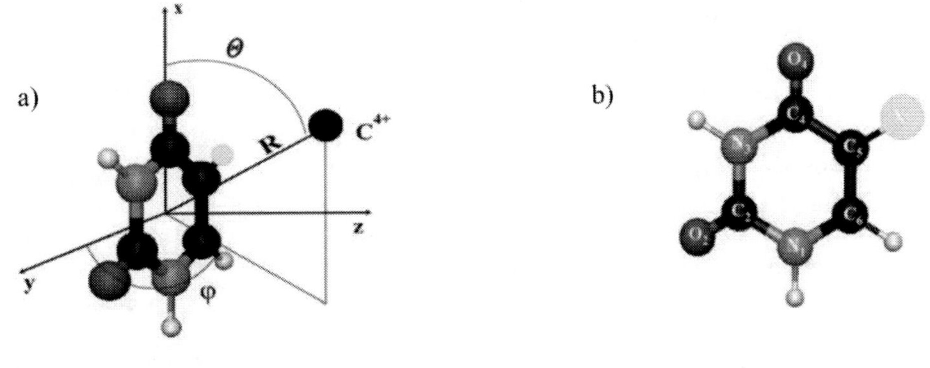

Figure 1. (a) Internal coordinates for the C^{4+} + biomolecule system. (b) Geometry of the biomolecule: thymine, X=CH_3; uracil, X=H; 5-fluorouracil, X=F; 5-chlorouracil, X=Cl; 5-bromouracil, X=Br.

The molecular calculations have been carried out using the MOLPRO suite of *ab-initio* programs [30]. As spin-orbit coupling is negligible in the energy range of interest, only singlet states have been considered in the calculation. An all-electron calculation has been performed with no symmetries for non-planar geometries using Cartesian coordinates with the origin of coordinates at the centre-of-mass of the target molecule. For the planar approach, calculations have been performed in the C_s symmetry group. Only the states coupled by means of radial coupling matrix elements have been taken into account in the calculation, the rotational coupling being neglected in a first step [31]. The potential energies and non-adiabatic coupling matrix elements (NACME) have been determined by state-averaged CASSCF calculations. Even if dynamic correlation effects are not taken into account at this level of theory, a correct description of the relative energies of the different excited states can be expected. Besides, some calculations at the higher MRCI (Multireference Configuration Interaction) level of theory have been performed for a few critical points. A similar active space has been considered for the different targets. It includes the six highest valence orbitals constructed mainly on the $2p_z$ orbital on CH_3 for thymine, the $2p_z$, $3p_z$, $4p_z$ orbitals centered, respectively, on fluorine, chlorine, and bromine for halouracil molecules, and, in all cases, the $2p_z$ orbitals centred on the oxygen atoms, the $2p_{z(C5)}$ and $2p_{z(C6)}$ orbitals (describing the molecular orbital called by extension $\pi_{(C5C6)}$, see figure 1b), and the $2p_x$, $2p_y$ and $2p_z$ orbitals of the colliding carbon ion. The 1s orbitals of carbon, nitrogen and oxygen are treated as frozen core. The valence electrons have been described using the 6-311G** basis set of atomic orbitals for all atoms.

The charge transfer process is driven mainly by the non-adiabatic interactions at the vicinity of the avoided crossings [32,33] and corresponding non-adiabatic radial coupling matrix elements between all pairs of states of the same symmetry have been calculated numerically by means of the finite difference technique [34]:

$$g_{KL}(R) = \langle \psi_K | \partial / \partial R | \psi_L \rangle = \langle \psi_K(R) | \lim_{\Delta \to 0} \frac{1}{\Delta} | \psi_L(R+\Delta) - \psi_L(R) \rangle, \qquad (1)$$

which, taking account of the orthogonality of the eigenfunctions $|\psi_K(R)\rangle$ and $|\psi_L(R)\rangle$ for $K \neq L$ reduces to

$$g_{KL}(R) = \langle \psi_K | \partial / \partial R | \psi_L \rangle = \lim_{\Delta \to 0} \frac{1}{\Delta} \langle \psi_K(R) | \psi_L(R+\Delta) \rangle. \qquad (2)$$

The stability with regard to the differentiation step has been tested and the value $\Delta = 0.0012 a.u$ has been chosen [35] using the three-point numerical differentiation method for reasons of numerical accuracy. The centre of mass of the target has been taken as origin of electronic coordinates.

The collision dynamics has been treated from keV to eV energies using the EIKONXS program based on an efficient propagation method [36] taking into account of all the transitions driven by radial coupling matrix elements. The calculations have been performed in the framework of the sudden approximation hypothesis assuming that electronic transitions occur so fast that vibration and rotation motions remain unchanged. Effectively the collision time to travel a distance L, $T_{coll} \approx L(\mu / 2E_{coll})^{1/2}$ [37], appears much shorter than the typical vibration time $T_{vib} \approx \Delta E_{vib}^{-1}$, and *a fortiori* typical rotation time. The ratio T_{vib} / T_{coll} has been shown to be about 10^2 at keV collision energies [12]. It depends of course of the collision velocity and decreases by a factor 10 between 1keV and 10 eV, remaining still in the approximation limit. The total and partial cross sections, corresponding to purely electronic transitions, are thus determined by solving the impact-parameter equation as in the usual ion-atom approach, considering the geometry of the molecular target fixed [29]. Such a treatment is, of course, relatively crude, but it has proved its efficiency in a number of ion-diatomic or polyatomic collisions [38,39,11] for energies higher than ~10eV/amu. Recently we have extended this approach to lower collision energies [20,22] with regard to our comparative analysis of time-dependent quantal wave packet and semiclassical approaches in charge transfer processes showing that semiclassical methods can give quite reasonable cross sections up to 10-20 eV, the discrepancy with quantal calculations appearing only for lower energies [40].

3. Anisotropic effect

For all biomolecular targets, the charge transfer process appears to be highly anisotropic. The cross sections calculated for a series of angles θ (see figure 1a) corresponding to different orientations of the projectile towards the target from perpendicular to planar geometries, show indeed very strong variations, as presented in figure 2 for the C^{4+} + uracil charge transfer system. For this system, the charge transfer is clearly favoured for geometries close to the perpendicular orientation. The cross sections are maximum for the angle θ=70°, about 2 orders of magnitude higher than in the planar or near-planar orientation. This result is observed for collision energies in the keV range as presented in figure 2a, but also at lower collision energies, down to 30 eV, as exhibited in figure 2b. Globally speaking, the charge transfer process appears thus clearly more efficient for the C^{4+} + uracil collision in a solid angle about 20° around the perpendicular geometry.

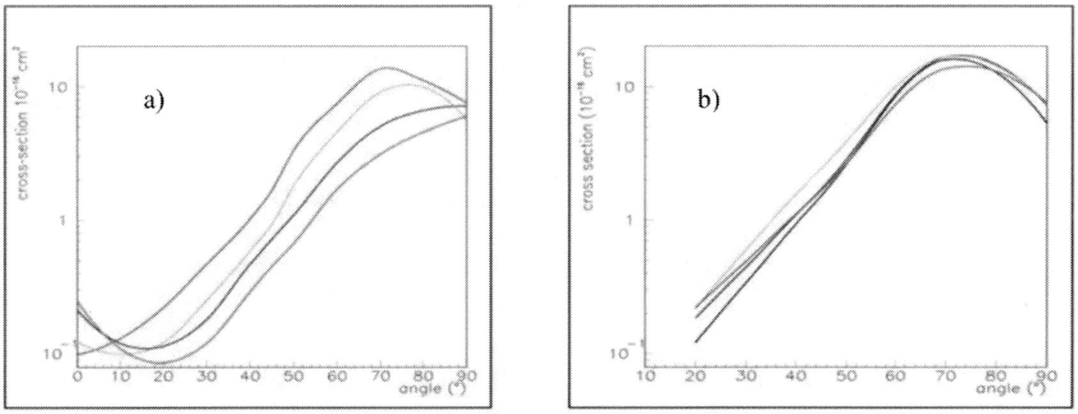

Figure 2. Charge transfer cross sections (given in 10^{-16} cm^2) for the C^{4+} + uracil collision system. (a) —, E_{lab}=3 keV; , E_{lab}=27 keV; —, E_{lab}=75 keV; —, E_{lab}=147 keV; (b) . —, E_{lab}=3 keV; , E_{lab}=750 eV; —, E_{lab}=120 eV; —, E_{lab}=30 eV.

Such effect is also observed for the halouracil targets. The results for collision energies between 3 keV and 30 eV are presented in figure 3a,b,c for, respectively, the 5-fluoro, 5-chloro and 5-bromouracil targets. The charge transfer process is clearly favoured near the perpendicular direction for the C^{4+} + 5-fluorouracil collision, with a maximum around the angle θ=70°. On the contrary, the process appears more efficient for an orientation around θ=45° for 5-chlorouracil or 5-bromouracil targets. A similar behaviour appears also at higher energies, in the keV energy range [12] and could certainly be attributed to the influence of a strong steric effect. The atomic radius of fluor is indeed small and cannot hinder the access of the projectile ion, leading to a mechanism quite similar to the one observed for the C^{4+} + uracil system with a preferred collision around the perpendicular geometry. This conclusion can also be extended to the thymine target molecule [22]. On the other hand, for a heavy substituent as chlorine or bromine of higher atomic radius, the access of the colliding ion in the perpendicular geometry is reduced and thus an approach further from the heavy halogen atom is favoured, as shown by the preferential cross section values in the θ=45° geometry.

Some specific behaviour can be pointed out for the C^{4+} + thymine charge transfer at eV collision energies. The corresponding cross sections in the [7.5-480] eV energy range are presented in figure 4 and show clearly a strong decrease of the charge transfer cross sections for the lowest energy, E_{lab}=7.5 eV. This effect is quite significant as charge transfer cross sections are lowered by a factor up to 10^2 at eV energies compared to cross sections at 500 eV. The effect is reduced with increasing energy, still marked at E_{lab}=30 eV, it disappears for collisions energies of the order of 100-500 eV. The cross sections appear to be highly anisotropic with a maximum for the orientation angle θ=20°-30°. This

Journal of Physics: Conference Series **373** (2012) 012002 doi:10.1088/1742-6596/373/1/012002

result may be discussed with regard to a previous analysis of the C^{q+} + uracil collision system involving the theoretical determination of charge transfer cross sections [11] and the experimental fragmentation yield [7]. In collisions of ions with biomolecular targets, excitation, ionization, fragmentation and charge transfer are strongly related processes and depend on a great number of factors, velocity, projectile charge, target state, but a qualitative correlation may be pointed out between fragmentation and charge transfer. For example, at low collision energy, the C^{2+} + uracil system appears experimentally to drive almost complete fragmentation, in relation with a very low charge transfer cross section and the fragmentation yield decreases with increasing collision energy, when the charge transfer cross section increases [7,11]. Conversely, for the same reactions with the C^{4+} projectile ion, the experimental fragmentation yield is lower, almost independent of the collision energy, and the calculated charge transfer cross sections are significantly higher than in the collision with the C^{2+} projectile, and similarly almost independent of the collision energy. These observations remain of course completely qualitative and have to be handled with care; anyway they would suggest that a correlation could be established between both processes. Considering the present results, fragmentation could thus be enhanced in the collision of C^{4+} ions on thymine for specific orientations at very low collision energies. This result is in accordance with experimental studies at hyperthermal energies [18,19] pointing out a complex mechanism in ion-induced radiation damage. This result has to be brought together with a similar enhancement observed for collisions with very low-energy electrons [3,4] and could be of interest for ion induced radiation damage.

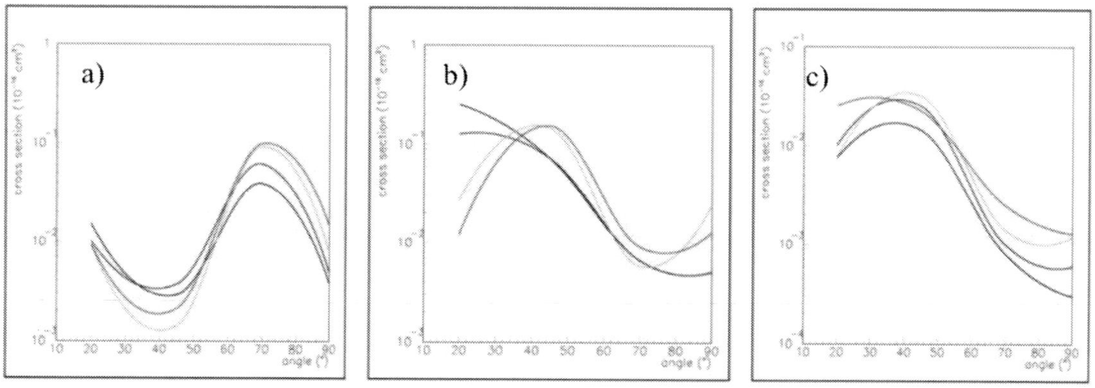

Figure 3. Charge transfer cross sections with regard to the orientation angle θ (given in 10^{-16} cm^2). (a) C^{4+}+ 5-fluorouracil; (b) C^{4+}+ 5-chlorouracil; (c) C^{4+}+ 5-bromouracil. —, E_{lab}=3 keV; , E_{lab}=750 eV; —, E_{lab}=120 eV; —, E_{lab}=30 eV.

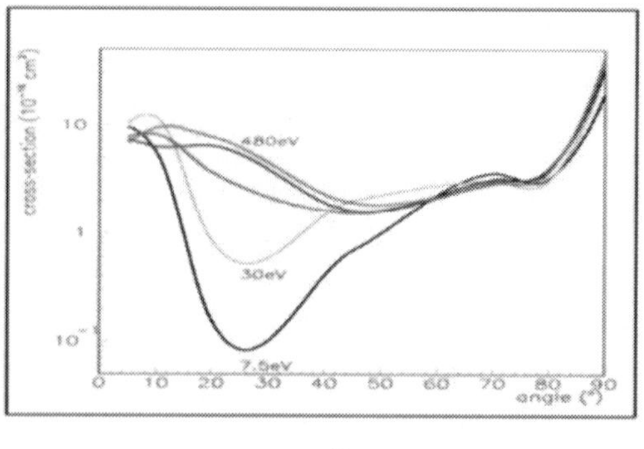

Figure 4. Charge transfer cross sections for the C^{4+}+ thymine system at low collision energies as a function of the θ angle (in 10^{-16} cm^2). —, E_{lab}=7.5 eV; , E_{lab}=30 eV; —, E_{lab}=120 eV; —, E_{lab}=270 eV; —, E_{lab}=480 eV.

Journal of Physics: Conference Series **373** (2012) 012002 doi:10.1088/1742-6596/373/1/012002

4. Radio-sensitivity properties

Independently of the anisotropic effect discussed in the previous paragraph, charge transfer cross sections have also to be analyzed with regard to their order of magnitude for the different biomolecular targets. The results are summarised in table 1 for a wide range of collision energies.

Table 1. Charge transfer cross sections averaged over the different orientations for a series of C^{4+}-biomlecule collision systems (in 10^{-16} cm^2).

Elab (eV)	C^{4+}+thymine [22]	C^{4+}+uracil [20,22]	C^{4+}+ fluorouracil [20,22,15]	C^{4+}+ chlorouracil [20,22,15]	C^{4+}+ bromouracil [20,22,15]
30	9.10	5.73	0.015	0.085	0.006
120	8.68	6.45	0.020	0.072	0.009
270	9.01	7.74	0.029	0.056	0.008
$3.\times10^3$	14.93	5.79	0.031	0.047	0.013
$27.\times10^3$	12.12	3.97	0.018	0.025	0.004

The cross sections are of the same order of magnitude for the thymine and uracil targets; the values are slightly higher for thymine with a maximum of 14.9 10^{-16} cm^2 whereas the charge transfer cross sections do not exceed 7.7 10^{-16} cm^2 for uracil. From our previous discussion, this supposes that fragmentation would be somewhat less efficient for collisions with the thymine target than observed for uracil. However, the effect would remain relatively smooth, compared with the completely different order of magnitude observed for the charge transfer cross sections involving halouracil targets. The corresponding cross sections are indeed at least a factor 100 lower than previously noticed for uracil or thymine targets. The rate reaches even up to 10^3 for 5-bromouracil. This point is crucial with regard to the radio-sensitivity of halouracils. Effectively, 5-bromouracil has been shown early to enhance DNA damage [25], but such sensitivity to ionizing radiation is also recognized for other 5-halouracil molecules and widely used in radiation therapy [41,42]. The very low values obtained for the charge transfer cross sections with halouracils would induce a very high efficiency for fragmentation, in complete agreement with the enhancement of DNA damage observed with 5-halouracils. Even if our approach remains qualitative, it could provide an order of magnitude of the effect. Effectively, the cross sections are numerically lower for the 5-bromouracil target supposed to be the most efficient radio-sensitizer. Such specific behaviour of halouracils could be attributed to an important electronic effect. Halogen atoms present indeed a high electro-negativity compared to the corresponding substituent, hydrogen atom or the CH_3 group, present in uracil or thymine targets which could compete with the electronic transfer from the target to the colliding carbon cation and charge transfer would be then less efficient.

5. Concluding remarks

The charge transfer in collisions of C^{4+} ions with thymine, uracil and halouracil, has been investigated using *ab-initio* methods in the one-dimension reaction coordinate approximation followed by a semiclassical collision treatment in the framework of the sudden approximation approach. Such approach provides information on the molecular states involved in the process, considering neutral and ionized target levels. The charge transfer process has been shown to be highly anisotropic, clearly favoured around the perpendicular geometry for thymine, uracil and fluorouracil, but quite sensible to steric effects as shown for chloro- and bromouracil targets for which the charge transfer is favoured in an orientation opposite to the heavy atom. A comparison with experimental measurements at given orientations would be highly desirable. Besides, a specific decrease of the C^{4+} + thymine charge transfer cross sections at eV energies, in an orientation close to the planar geometry, has been exhibited. This could lead to an interesting enhancement of the fragmentation process at very low energies, as observed for collisions with low-energy electrons and further investigations would be welcome. Taking into account the correlation pointed out between charge transfer and fragmentation processes, a strong radio-sensitization effect may be pointed out for the halouracil targets, in particular the 5-bromouracil, which would induce an enhancement of DNA damage through ionizing radiation in agreement with medical observations.

Acknowledgments
We acknowledge support from the HPC resources of CCRT/CINES/IDRIS under the allocation 2011-[i2011081566] made by GENCI [Grand Equipement National de Calcul Intensif] as well as from the COST action MP1002 Nano-IBCT.

References
[1] von Sonntag C 1987 *The Chemical Basis for Radiation Biology* (Taylor and Francis, London)
[2] Michael B D and O'Neill P D 2000 *Science* **287** 1603
[3] Boudaiffa B, Cloutier P, Hunting D, Huels M A and Sanche L 2000 *Science* **287** 1658
[4] Pan X, Cloutier P, Hunting D and Sanche L 2003 *Phys. Rev. Lett.* **90** 208102
[5] Martin F, Burrow P D, Cai Z, Cloutier P, Hunting D and Sanche L 2004 *Phys. Rev. Lett.* **93** 068101
[6] Abdoul-Carime H, Huels M A, Illenberger E and Sanche L 2001 *J. Am. Chem. Soc.* **123** 5354
[7] de Vries J, Hoekstra R, Morgenstern R and Schlathölter T 2002 *J. Phys.* B **35** 4373
[8] de Vries J, Hoekstra R, Morgenstern R and Schlathölter T 2003 *Phys. Rev. Lett.* **91** 053401
[9] Coupier B, Farizon B, Farizon M, Gaillard M J, Gobet F, de Castro Faria N V, Jalbert G, Ouaskit S, Carré M, Gstir B, Hanel G, Denifl S, Feketeova L, Scheier P and Märk T D 2002 *Eur. Phys. J.* D **20** 459
[10] de Vries J, Hoekstra R, Morgenstern R and Schlathölter T 2003 *Eur. Phys. J.* D **24** 161
[11] Bacchus-Montabonel M C, Łabuda M, Tergiman Y S and Sienkiewicz J E 2005 *Phys. Rev.* A **72** 052706
[12] Bacchus-Montabonel M C andTergiman Y S 2006 *Phys. Rev.* A **74** 054702
[13] Alvarado F, Bari S, Hoekstra R and Schlathölter T 2006 *Phys. Chem. Chem. Phys.* **8** 1922
[14] Schalhölter T, Alvarado F, Bari S, Lecointre A, Hoekstra R, Bernigaud V, Manil B, Rangama J and Huber B A 2006 *ChemPhysChem* **7** 2339
[15] Bacchus-Montabonel M C, Tergiman Y S and Talbi D 2009 *Phys. Rev.* A **79** 012710
[16] Bacchus-Montabonel M C, Tergiman Y S and Talbi D 2011 *Int. J. Quant. Chem.* **111** 520
[17] Lacombe S, Le Sech C and Esaulov V A 2004 *Phys. Med. Biol.* **49** N65
[18] Deng Z, Imhoff M and Huels M A 2005 *J. Chem. Phys.* **123** 144509
[19] Deng Z, Bald I, Illenberger E and Huels M A 2005 *Phys. Rev. Lett.* **95** 153201
[20] Bacchus-Montabonel M C and Tergiman Y S 2011 *Chem. Phys. Lett.* **503**, 45
[21] Bacchus-Montabonel M C and Tergiman Y S 2010 *Chem. Phys. Lett.* **497** 18
[22] Bacchus-Montabonel M C and Tergiman Y S 2011 *Phys. Chem. Chem. Phys.* **13** 9761
[23] Illescas C, Errea L F, Méndez L, Pons B, Rabadán I and Riera A 2011 *Phys. Rev.* A **83** 052704
[24] Bene E, Vibók Á, Halász G J and Bacchus-Montabonel M C 2008 *Chem. Phys. Lett.* **455** 159
[25] Bacchus-Montabonel M C 2012 *Eur. Phys. J. D* (in press)
[26] Zamenhof S, DeGiovanni R and Greer S 1959 *Nature* (London) **181** 827
[27] Salem L 1982 *Electrons in Chemical Reactions: First Principles* (Wiley Interscience, New York)
[28] Bacchus-Montabonel M C, Talbi D and Persico M 2000 *J. Phys.* B **33** 955
[29] Errea L F, Gorfinkiel J D, Harel C, Jouin H, Macías A, Méndez L, Pons B and Riera A 1996 *Phys. Scr.* **T62** 33
[30] Werner H J and Knowles P J (2010) *MOLPRO package of* ab initio *programs*

[31] Honvault P, Bacchus-Montabonel M C and McCarroll R 1994 *J. Phys.* B **27** 3115
[32] Bacchus-Montabonel M C, Vaeck N, Lasorne B and Desouter-Lecomte M 2003 *Chem. Phys. Lett.* **374** 307
[33] Gargaud M, Fraija F, Bacchus-Montabonel M C and McCarroll R 1994 *J. Phys.* B **27** 3985
[34] Bacchus-Montabonel M C, Cimiraglia R and Persico M 1994 *J. Phys.* B **17** 3985
[35] Bacchus-Montabonel M C 1992 *Phys. Rev.* A **46** 217
[36] Allan R J, Courbin C, Salas P and Wahnon P 1990 *J. Phys.* B **23** L461
[37] Sidis V 1990 *Adv. At. Mol. Opt. Phys.* **26** 161
[38] Stancil P C, Zygelman B and Kirby K 1998 *Photonic, Electronic, and Atomic Collisions* (World Scientific, Singapore)
[39] Bene E, Martínez P, Halász G J, Vibók Á and Bacchus-Montabonel M C 2009 *Phys. Rev.* A **80** 012711
[40] Chenel A, Mangaud E, Justum Y, Talbi D, Bacchus-Montabonel M C and Desouter-Lecomte M 2010 *J. Phys.* B **43** 245701
[41] Lawrence T S, Davis M A, Maybaum J, Stetson P L and Ensminger W D 1990 *Radiat. Res.* **123** 192
[42] McLaughlin P, Mancini W, Stetson P, Greenberg H, Nguyen N, Seabury H, Heidorn D and Lawrence T S 1993 *Int. J. Radiat. Oncol., Biol. Phys.* **26** 637

Quantum mechanical calculations related to ionization and charge transfer in DNA

E Cauët[1], M Valiev[2], J H Weare[3] and J Liévin[1]

[1] Service de Chimie Quantique et Photophysique, Université Libre de Bruxelles, 50 Av. F.D. Roosevelt, CP 160/09, B-1050 Brussels, Belgium
[2] Pacific Northwest National Laboratory, MS K8-91, P.O. Box 999, Richland, WA 99352, USA
[3] Chemistry and Biochemistry Department, University of California, San Diego, 9500 Gilman Dr., La Jolla, CA 92093, USA

Email: ecauet@ulb.ac.be

Abstract. Ionization and charge migration in DNA play crucial roles in mechanisms of DNA damage caused by ionizing radiation, oxidizing agents and photo-irradiation. Therefore, an evaluation of the ionization properties of the DNA bases is central to the full interpretation and understanding of the elementary reactive processes that occur at the molecular level during the initial exposure and afterwards. *Ab initio* quantum mechanical (QM) methods have been successful in providing highly accurate evaluations of key parameters, such as ionization energies (IE) of DNA bases. Hence, in this study, we performed high-level QM calculations to characterize the molecular energy levels and potential energy surfaces, which shed light on ionization and charge migration between DNA bases. In particular, we examined the IEs of guanine, the most easily oxidized base, isolated and embedded in base clusters, and investigated the mechanism of charge migration over two and three stacked guanines. The IE of guanine in the human telomere sequence has also been evaluated. We report a simple molecular orbital analysis to explain how modifications in the base sequence are expected to change the efficiency of the sequence as a hole trap. Finally, the application of a hybrid approach combining quantum mechanics with molecular mechanics brings an interesting discussion as to how the native aqueous DNA environment affects the IE threshold of nucleobases.

1. Introduction

Ionization via high-energy radiation of the components of DNA and charge transfer in DNA are key processes in the propagation and detection of DNA radiation damage occurring in the cells of human beings [1]. As we know now, DNA radiation damage is responsible for many neurological diseases and plays an important role in aging and many forms of human cancer. Each portion of the DNA molecule and its environment such as the nucleobases, the sugar-phosphate backbone and the water molecules of hydration are ionized in a near random fashion. However, owing to charge transfer, the chemical damage in irradiated DNA often occurs at sites other than where the original ionization takes place. For these reasons, the evaluation of the ionization properties of the components of DNA as well as the elucidation of the mechanisms responsible for the charge transfer phenomena in DNA are a great challenge in the current research on the oxidative damage of DNA. This challenging problem is also essential in the nanoscience theme area [2]. In particular, the ability of DNA to serve as a medium

Published under licence by IOP Publishing Ltd

for long-range charge transfer has stimulated interest in the possibility to exploit this molecule in nanoscale electronics and in electrochemical biosensor electronics.

The improvement of quantum calculation techniques, together with the increase of computing power, makes the quantum approach particularly suitable to study, on the molecular scale, elementary mechanisms related to the DNA damage process [3]. Thus, the aim of our study was to apply high level methods of quantum chemistry in a complementary way investigating several aspects of this problem such as the ionization of the DNA bases, isolated and embedded in base clusters, the sensitivity of the sequence, in particular the human telomere overhang sequence, to oxidation and the role of structural variations in the efficiency of charge transport in DNA. In addition, particular attention has been given to the effects of the DNA environment structure and dynamics on the energetic of the ionization process generated in DNA.

2. Computational Methods

Various sophisticated "post Hartree-Fock" methods considering explicitly that electron motions are correlated, have been used in this work. These methods are based on the three main wave function based approaches for calculating electron correlation: Many Body Perturbation Theory (MBPT), Coupled Cluster (CC) and Configuration Interaction (CI). The Density Functional Theory (DFT), the alternative quantum mechanical electron correlation method in which the total energy is expressed in terms of the total electron density rather than the wave function, has also been used. Concerning the choice of the basis set, the 6-31G(2d(0.8,0.2),p) basis set, proposed by our group [4], which corresponds to the standard 6-31G(d,p) basis set augmented by diffuse polarization functions (exponent $\alpha_d = 0.2$) on the heavy atoms, has mainly been exploited for investigating ionization of large systems involving the DNA bases. Indeed, it has been demonstrated that this basis set can be used to predict the H-bond and π-π energies and to compute the energy differences related to ionization as well as larger polarized and/or augmented basis sets of the literature but at lower computer costs. Thanks to this optimized basis set, the study of large ionized clusters related to DNA has been made possible.

3. Results and Discussion

3.1. Ionization energy of stacked DNA bases

While the ionization of isolated DNA bases has been studied extensively, only a few *ab initio* ionization energy (IE) calculations on π-stacked bases have been reported. Indeed, whereas the qualitative features of ionization of stacked systems, i.e. the lowering of IE and the dependence of hole delocalization on the relative orientation of the fragments, can be easily explained by an electrostatic model and simple molecular orbital analysis [5], the quantitative predictions are much more challenging. In our study, the influence of stacking interactions on IE of guanine (G), guanine being the most easily oxidized nucleic acid base, has been examined by employing pure guanine single-stranded model systems [6]. These consist of stacked bases in orientations that occur in the standard B form of DNA. In this case, the geometries of the stacked systems were not optimized and the calculated values are vertical IEs. Table 1 summarizes the results, which include MP2 IE estimates using the 6-31G(2d(0.8,0.2),p) basis set. Stacking interactions in the GG and GGG sequences are found to reduce the vertical IE by 0.4-0.5 eV. The IE lowering by stacking interactions is further confirmed for four, five and six stacked guanines, in which the IE gradually drops with increasing number of nucleobases.

Table 1. Vertical IEs (in eV) of the B-form stacked contiguous guanines. Adapted from Cauët [6].

5'-Base(s)-3'	MP2/6-31G(2d,p)
G	8.07
GG	7.67
GGG	7.54
GGGG	6.92
GGGGG	6.81
GGGGGG	6.57

3.2. Human telomere sequence

The hypothetical protection of genes from oxidative damage provided by the G-rich telomeric overhangs located at the end of chromosomes, which consist, in humans, of single strands of TTAGGG sequence repeats, has been investigated [6]. First principle MP2 calculations reveal that the TTAGGG human telomere sequence is particularly prone to oxidation (IE=6.76 eV) and can act as a profound hole trap as deep as a sequence of five consecutive guanines (see table 1). In addition, IEs were estimated, at the same level of calculation, for TTGGGG (7.30 eV), GGGATT (7.48 eV) and GGGGTT (7.39 eV) sequences. These show that the sequence dependence is very important and that modifications in the human telomeric sequence can induce crucial changes in the electronic structure of the sequence, with concomitant increase of the ionization energy. Interestingly, the expected most electron-releasing 5' side G in the TTGGGG sequence is not susceptible to oxidation in contrast to the human telomeric sequence TTAGGG in which the 5' side G is found to be efficiently oxidized (see figure 1). According to our calculations, this is caused by the fact that the molecular orbitals which are localized on the 5' side G are largely stabilized in the cationic state of TTGGGG due to an orbital mixing with the orbitals from the thymine bases. The molecular orbitals of the cationic state of the human telomeric sequence, however, show that no orbital mixing appears due to the presence of A.

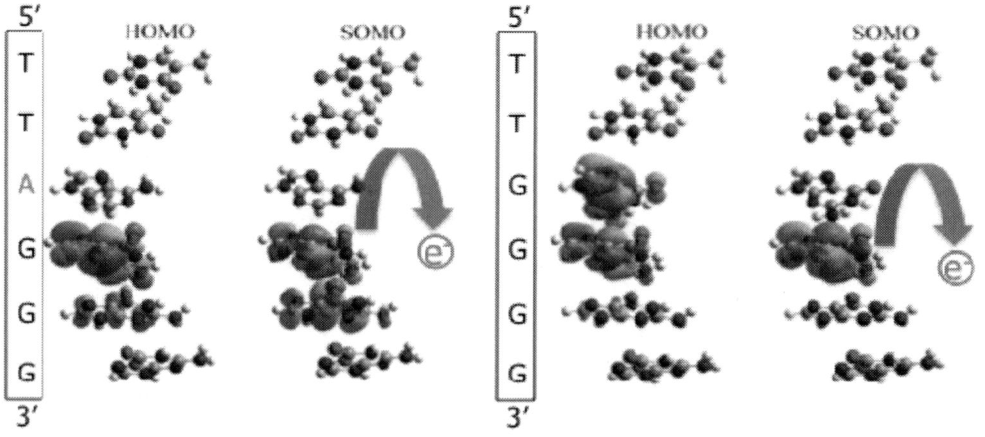

Figure 1. Isosurfaces of the π-like HOMO and singly occupied HOMO (SOMO) orbitals of the neutral and ionized species, respectively, calculated at the HF/6-31G(2d(0.8,0.2), p) level, for the 5'-TTAGGG-3' and 5'-TTGGGG-3' sequences. The two different colours of the isosurfaces correspond to the different signs of the wave function. Adapted from Cauët [6].

3.3. Charge Transfer in π-stacked guanine clusters

High-level multiconfigurational methods have been applied to determine the reaction path describing the charge transfer between two consecutive stacked guanines [7]. From these calculations, the most important conformational changes that accompany the charge transfer process have been identified as a slide and a shift of one base with respect to the other. Figure 2 presents the topology of the ground-state potential energy surface of a radical cation $(G_{5'}G_{3'})^+$ calculated at the RASSCF/RASPT2/6-31G(d(0.2)) level of theory as a function of the parameters *q* and *l* describing translation motions of $G_{5'}$ along the axes q (slide) and l (shift), respectively. The topology shows five stationary points: three minima, denoted M1, M2 and M3, and two saddle points, S1 and S2. The minimum M1, in which most of the positive charge is on $G_{5'}$ is more stable by 3.84 kcal/mol than M3 where most of the charge is on $G_{3'}$. The intermediate minimum M2, where the positive charge is almost equally distributed between the two guanines, is destabilized by 2.13 kcal/mol relative to M1. Furthermore, our study has been extended to a stacked sequence of three guanines. The results of the calculations performed on these trimer geometries are promising and exciting as they show that the charge transfer proceeds in a multi-step fashion, from one base to another, across the trimer following the mechanism we have predicted from the dimer study.

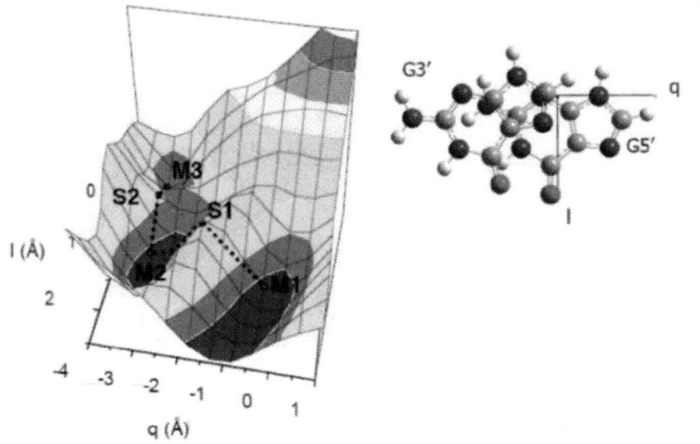

Figure 2. Energy map for the ground state calculated at the RASSCF/RASPT2/6-31G(d(0.2)) level for the radical cation $(G_{52}G_{32})^+$ as a function of the parameters *q* and *l*. The vertical separation between the two stacked guanines is kept fixed equal to 3.3 Å. Five stationary points have been localized (M for minimum, S for saddle point). The point (0,0) corresponds to the reference crystal structure 1TC3. Adapted from Cauët *et al.* [7].

3.4. Influence of the native DNA environment

As described in the earlier sections, quantum mechanics calculations may be applied for systems where no relevant experiments exist, such as stacked and H-bonded nucleic acid bases. However, it is important to recognize that in physiological surroundings the nucleobases interact with solvent, counterions and the sugar-phosphate backbone of the double-stranded DNA helix. All these effects can cause important changes in the electronic structure of the DNA bases and the next step is thus to ask what are the IEs of DNA bases if we take these into account. We have recently applied a hybrid quantum mechanics/molecular mechanics (QM/MM) methodology at the CCSD(T)/aug-cc-pVDZ level of theory to estimate the vertical IEs of the four individual DNA bases embedded in a 12-mer fragment of B-DNA neutralized by 22 sodium cations and solvated with 5760 water molecules [8]. The results are collected in table 2 where they are compared to the gas-phase IE values computed at the same level. The large increase of the vertical IEs in the hydrated DNA environment may, at first, seem surprising given the fact that hydration is expected to screen electrostatic interactions and thus stabilize the cationic species whereas the opposite is observed here. However, the latter notion is for bulk and it does not apply to the special hydration process of DNA. Indeed, the solvent structure created by the anionic phosphates and counterions intensifies the attractive interaction between the electrons and the dipolar charge distribution of the water molecules increasing thus the energy necessary to ionize the bases.

Table 2. Results of gas-phase and QM/MM calculations of vertical ionization energies (in eV) of guanine, adenine, cytosine and thymine. Energies (Δ) of QM/MM results relative to the calculated gas-phase values are included. Adapted from Cauët *et al.* [8].

CCSD(T)/aug-cc-pVDZ	Gua	Ade	Cyt	Thy
Gas phase	7.93	8.23	8.65	8.98
QM/MM	11.09	11.55	11.98	12.33
Δ	3.16	3.32	3.33	3.35

4. Conclusions

The present findings point to interesting phenomena in which stacking interactions and long-range modulation of the solvent structure strongly affect the IEs of DNA bases. Despite these considerable efforts, continuous work to provide accurate theoretical values of the IEs of the DNA bases with reliable methods and proper environment representation is still required. Several important issues remain to be further explored such as the sequence dependence of IE and the influence of geometry relaxation. Indeed, consideration of geometry relaxation of a stacked cluster may be expected to induce changes of the IE of the stack that are of the same order of magnitude as the ones discussed in this work and would allow for a more quantitative treatment of the ionization of these complexes.

References

[1] *Charge Migration in DNA. Perspectives from Physics, Chemistry and Biology.* 2007 Chakraborty, Tapash (Ed.)
Merino E J, Boal A K and Barton J K 2008 *Curr. Opin. Chem. Biol.* **12** 229
[2] Porath D, Cuniberti G and Di Felice R 2004 *Top. Curr. Chem.* **237** 183
Guo X, Gorodetsky A A, Hone J, Barton J K and Nuckolls C 2008 *Nat. Nanotechnol.* **3** 163
[3] Cauët E *Quantum mechanicals methods related to ionization of nucleic acid bases,* Chapter 5 in *DNA Microarrays, Synthesis and Synthetic DNA.* 2011 Marissa J Campbell (Ed), Nova Science Publishers.
Cauët E and Liévin J 2007 *Adv. Quantum Chem.* **52** 121
[4] Cauët E, Dehareng D and Liévin J 2006 *J. Phys. Chem.* A **110** 9200
Wintjens R, Biot C, Rooman M and Liévin J 2003 *J. Phys. Chem.* A **107** 6249
[5] Prat F, Houk K N and Foote C S 1998 *J. Am. Chem. Soc.* **120** 845
Saito I, Nakamura T and Nakatani K 2000 *J. Am. Chem. Soc.* **122** 3001
[6] Cauët E 2011 *J. BiolMol. Struct. Dyn.* **29** 557
[7] Cauët E and Liévin J 2009 *J. Phys. Chem.* A **113** 9881
Rooman M, Cauët E, Liévin J and Wintjens R 2011 *J. BiolMol. Struct. Dyn.* **28** 949
[8] Cauët E, Valiev M and Weare J H 2010 *J. Phys. Chem.* A **114** 5886

Nano-IBCT 2011 IOP Publishing
Journal of Physics: Conference Series **373** (2012) 012004 doi:10.1088/1742-6596/373/1/012004

Ionization of RNA-uracil by highly charged carbon ions

C Champion[1], M E Galassi[2], O Fojón[2], H Lekadir[1], J Hanssen[1], R D Rivarola[2], P F Weck[3], A N Agnihotri[4], S Nandi[4], and L C Tribedi[4]

[1] Laboratoire de Physique Moléculaire et des Collisions, UMR CNRS 7565, Université de Lorraine, Metz, France
[2] Instituto de Física Rosario, CONICET and Universidad Nacional de Rosario, Argentina
[3] Department of Chemistry and Harry Reid Center for Environmental Studies, University of Nevada Las Vegas, USA
[4] Tata Institute of Fundamental research, Colaba, Mumbai, India

Email: champion@univ-metz.fr

Abstract. We present a theoretical description of highly charged carbon ion-induced ionization of isolated RNA-uracil molecules. A comparison between recent experimental and theoretical total cross sections is provided.

1. Introduction

Besides their great importance in fundamental physics, atomic and molecular ionization processes are of prime interest in diverse areas of science ranging from astrophysics - where the effects of space radiation on astronauts is an important concern of missions for the human exploration of the solar system - to radiobiology, in particular for predicting the radio-induced damage at the sub-cellular scale [1]. Indeed, it is nowadays recognized that lethal events for living cells are - in their majority - attributed to complex clustered damages coupling DNA base lesions with single- and double-strand breaks [2]. Thus, the ionization process being known to be one of the main elicitors of these double strand breaks (DSBs), it is worth noting that a detailed knowledge of the underlying physics of the ionizing radiation remains of prime importance for radiobiologists. In this context, accurate cross sections for describing the ionization of biomolecules like the DNA/RNA components represent useful data, in particular for Monte Carlo type numerical simulations devoted to the modeling of radio-induced cellular damage [3-4]. Additionally, knowledge of the energy deposition patterns in the biological medium is of crucial importance in medicine and more particularly in cancer therapy where focused ion beams are used to deliver highly localized doses within tumors while sparing the surrounding normal tissue. This particular ballistic property may only be achieved by highly charged particles which offer the possibility to concentrate the radiation dose in a well defined depth called Bragg peak (BP). Besides, it has been shown that the ratio of BP versus dose in the entrance region is larger for heavy ions than for protons. Indeed, due to their larger mass, angular and energy straggling becomes negligible for heavy ions as compared to protons. Therefore, heavy ions offer an improved dose conformation as compared to photon and proton radiation therapy, with a better sparing of normal tissue structures close to the target. Consequently, if the irradiated volume of normal tissue can be minimized by conformal radiation therapy, a higher dose can be delivered to the tumor and thus a better outcome can be achieved without increasing the risk of side effects. This property is the basis of most developments in the field of radiation therapy in the last decades, the highest degree of dose conformation being currently achieved with protons and carbon ions. Moreover, in addition to the dose conformation, heavy ions exhibit a strong increase of the Linear Energy Transfer (LET) in the BP as compared to the entrance region, leading to a radiobiological advantage of high LET radiation versus neutron therapy.

However, investigations concerning proton- and heavy ion-induced ionization of DNA/RNA constituents remain limited. On the experimental side, Tabet *et al.* [5-6] have measured proton-impact total ionization cross sections of nucleobases such as adenine and uracil by means of a time-of-flight fragment ion measurements at proton energies below 150 keV. Previously, Moretto-Capelle and Le Padellec [7] have published energy distributions of secondary electrons emitted from uracil by protons with energies below 100 keV. Very recently, Iriki *et al.* have reported two extensive studies where a large set of absolute doubly and singly differential cross sections was provided for 500 keV, 1 MeV and 2 MeV protons impacting with adenine [8-9]. Finally, Agnihotri *et al.* have - in a recent work [10] - reported absolute total ionization cross sections for uracil molecules impacted by helium-like carbon and oxygen ions over a wide energy range (100 keV-60 MeV) which approximately covers the BP region. Detailed angular distributions in terms of doubly differential cross sections for MeV bare carbon ions on uracil were also studied by the same authors [11].

With regard to the theoretical work, let us first mention that still today they are relatively scarce. In this context, we have recently proposed a classical model combining a classical trajectory Monte Carlo (CTMC) approach with a classical over-the-barrier (COB) criterion. Herewith we described total cross sections for single and multiple ionizing processes induced by impact of fast H^+, He^{2+} and C^{6+} ions on RNA uracil and DNA bases [12-13]. Simultaneously, we have developed two quantum mechanical models, within the first-order Born approximation and the continuum distorted wave-eikonal initial state (CDW-EIS) framework, where ionization induced by protons on the four DNA bases was investigated [14-15]. These two models are referred to as CDW-EIS and CB1-CWB models, respectively. A detailed comparison between the CTMC-COB, CB1-CWB and CDW-EIS predictions was also reported in [10-11] where a good qualitative agreement was observed in terms of total and differential ionization cross sections for all the investigated cases.

In the current work, we present an extension of our previous work [14-15] to the case of C^{q+} ions (with q ranging from 4 to 6). The theoretical CB1-CWB and CDW-EIS predictions are compared with measured total ionization cross sections for isolated uracil ($C_4H_4N_2O_2$) molecules impacted by keV-MeV C^{q+} ions, *i.e.* for incident energies which largely cover the maximum of the Bragg Peak region.

In the following we briefly describe the experimental set-up built to measure total cross sections (TCS) for electron emission from uracil. In section 3 we briefly present the different theoretical models developed for describing the ionization process while section 4 contains a detailed comparison of theoretical and experimental TCS.

Note that in the following sections atomic units (a.u.) are used throughout unless indicated otherwise.

2. Experimental set-up

In this section we briefly report the experimental set-up used to measure the ionization cross sections; for more details we refer to our previous work [10-11,16].

The experiments were carried out by using two different accelerator facilities both based at the Tata Institute of Fundamental Research (Mumbai, India). Whereas low energy C-ion beams (60 keV-1 MeV) were delivered from a newly installed Electron Cyclotron Resonance ion-source, a 14 MV tandem Pelletron accelerator was used to produce high-energy carbon beams. A Wiley McLaren type recoil-ion time-of-flight spectrometer was used along with channel electron multipliers (CEM) to detect electrons and ions. A narrow slit was placed in front of the electron-CEM to reduce the excessive count rate.

The well-collimated ion beams were crossed with a gaseous uracil target in the interaction region between the pusher and puller plates, the molecules effusing through a nozzle of 1 mm diameter. The gaseous target of uracil was produced by heating the (99% pure, Sigma-Aldrich) powder in an oven at 160° C. The evaporation rate was monitored by means of a quartz crystal thickness monitor suitably mounted to monitor the flow of molecules throughout the experiments. Variation of the deposition rate, if at all, was very smooth and was found to vary by about 5-10% over a long period *i.e.* in about ten hours. The absolute error (~20%) includes this variation. Chamber pressure was better than 4×10^{-7} Torr. For absolute normalization of the TCS we have used another experimental technique based on electron spectroscopy, detailed in [11]. Total cross sections were then obtained by adding all the counts under the single ionization and fragmentation peaks (see Agnihotri *et al.* [10])

3. Theory

The different models used in the present study for predicting total cross sections of carbon-induced ionization of uracil have been presented in previous publications, so hereafter only a brief outline is presented; for more details we refer to our previous work [13-15].

3.1. The classical description of the ionization process

The model here employed is based on the classical over-the-barrier (COB) approximation - initially developed by Bárány *et al.* [17] and Niehaus [18] - combined to statistical features of the CTMC model introduced by Abrines and Percival [19].

In this approach, the collision process is described within the impact parameter approximation - with a projectile trajectory considered as a straight line (see figure 1)- and where all the particles are described by classical Newtonian laws. Under these conditions, the target position is considered as fixed in the laboratory frame whereas the active target electrons, namely, those which will be extracted from the target are, in a first step, considered as virtual particles and really taken into account only if they are created, *i.e.* if ionization and/or capture processes occur (see the criteria given below). In addition, the initial positions of the target as well as that of the ejected electron are fixed at the origin of the laboratory frame and then, step by step, modified via the Newtonian laws and according to the above mentioned criteria.

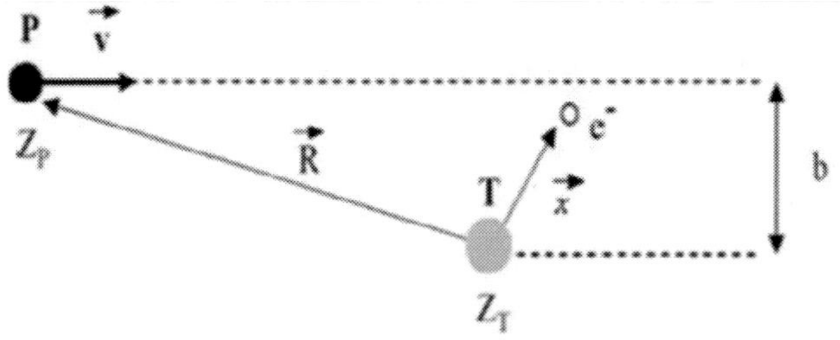

Figure 1. Geometry of a collision between a projectile P and a target T. Z_P and Z_T represent the charge of the projectile and the target, respectively, after the electron ejection. b and $\vec{v}$ represent the impact parameter and the impact velocity, respectively, whereas $\vec{x}$ is the position vector of the created electron and $\vec{R}$ is the internuclear vector.

At the beginning of each simulation, the system is only constituted by two particles, namely, the projectile nucleus P and the residual target T. Then, for each step Δt we evaluate the possibility for extracting one electron from the target by two simultaneous conditions: *i)* the maximum of the interaction potential seen by a target electron which has to be lower than the binding energy ε_α of the impacted orbital (ε_α <0) and *ii)* the ratio $\Delta t / t_e$, where t_e denotes the classical orbital period of the target electron given by $t_e = 2\pi Z_T / \sqrt{2|\varepsilon_\alpha|^3}$ with $\Delta t << t_e$, which has to be greater than a random number δ (chosen in the [0,1] interval). In fact, this last condition is essentially introduced to compensate the lack of target electron spatial density representation in the current model. Thus, if conditions *i)* and *ii)* are simultaneously satisfied, an electron is created and ejected from the target. The system becomes thus a three-free particle problem (the projectile, the residual target and the created electron).

At the end of the collision, the population of the different reacting channels is determined via energetic criteria. So, at the end of each simulation, *i.e.* at large values of R ($\cong$ 200 a.u.), the electron-projectile energy H_P and the electron-target energy H_T are intra-compared. Thus, if we have simultaneously $H_P > 0$ and $H_T > 0$ a single ionization (SI) process is considered while a single capture (SC) is considered if $H_P \leq 0$ or $H_T \leq 0$, respectively. The corresponding total cross section $\sigma_{SI,SC}$ is then simply determined by

$$\sigma_{SI,SC} = \sum_{i=1}^{I} \sigma^{i}_{SI,SC} = \sum_{i=1}^{I} 2\pi \int_{0}^{b_{\max}} b.db.P^{i}_{SI,SC}(b) \tag{1}$$

where the upper limit b_{max} of the integral has been found to be of the order of 15 a.u. for the system here studied (see [12]), whereas the *mono-electronic* probability $P^{i}_{SI,SC}(b)$ is simply defined by the ratio $P^{i}_{SI,SC}(b) = N^{i}_{SI,SC} / N_{tot}$ where $N^{i}_{SI,SC}$ corresponds to the number of SI or SC processes - occurring among N_{tot} ($\cong$ 500) simulations - for a given molecular orbital labelled i. Similar procedures are finally repeated for a large number ($\geq$ 100) of impact parameters.

Besides, I denotes the number of molecular subshells considered in the RNA-uracil description ($I = 21$). At this stage, it is important to note that in the current CTMC approximation, the biomolecule is simply described via its molecular orbital (MO) energies $\varepsilon_{\alpha} \equiv \varepsilon^{(i)}_{\alpha}$ in contrast to the hereafter developed quantum mechanical approaches which need an accurate description of the target in terms of molecular wave functions.

3.2. *The CDW-EIS and CB1-CWB descriptions of the ionization process*

Two different models are used to investigate the ionization reaction: the first-order Born approximation (denoted as CB1-CWB) and the continuum distorted wave-eikonal initial state one (CDW-EIS). The main difference between both models resides in the fact that in the CB1-CWB model, one-center wave functions (target and continuum states in the only presence of the residual target) are chosen, whereas in CDW-EIS, two-center wave functions (target and continuum states in the simultaneous presence of the projectile and residual target fields) are selected (for more details we refer the reader to [20]). For simplicity we describe here a mono-electronic target, being now Z_P and Z_T the projectile and nuclear charges, respectively.

Thus, in the CB1-CWB approach the initial and final wave functions are respectively given by the expressions

$$\chi^{+}_{\alpha} = \varphi_{\alpha}(\vec{x}) \exp(-i\varepsilon_{\alpha} t) \exp\left[-i \frac{Z_P}{v} \ln(vR - \vec{v} \cdot \vec{R}) \right] \tag{2}$$

and

$$\chi^{-}_{\beta} = (2\pi)^{-3/2} \exp(i\vec{k} \cdot \vec{x} - i\frac{k^2}{2} t) N^{*}(\xi)\, {}_1F_1(-i\xi; 1; -ikx - i\vec{k} \cdot \vec{x}) \exp\left[i \frac{Z_P}{v} \ln(vR + \vec{v} \cdot \vec{R}) \right] \tag{3}$$

where $\vec{v}$ is the collision velocity. The eikonal phases associated with the internuclear potential are neglected considering that they do not affect the calculation of cross sections integrated over the projectile scattering angle.

In equation (2) $\varphi_\alpha(\vec{x})$ represents a target bound state with corresponding orbital energy ε_α, whereas $\vec{k}$ (see equation (3)) denotes the linear momentum of the emitted electron as measured from the target nucleus and $\xi = Z_T/k$. $N(a) = \exp(\pi a/2)\Gamma(1+ia)$ refers to the normalization constant of the continuum factor ${}_1F_1(-i\xi;1;-ikx - i\vec{k}\cdot\vec{x})$.

Similarly, in the CDW-EIS framework, initial and final distorted wave functions are chosen as

$$\chi_\alpha^+ = \varphi_\alpha(\vec{x})\exp(-i\varepsilon_\alpha t)\exp\left[-i\frac{Z_P}{v}\ln(vs + \vec{v}.\vec{s})\right] \tag{4}$$

and

$$\chi_\beta^{-(i)} = (2\pi)^{-3/2}\exp(i\vec{k}.\vec{x} - i\frac{k^2}{2}t)N(\xi)\,{}_1F_1(-i\xi;1;-ikx - i\vec{k}.\vec{x})N(\zeta)\,{}_1F_1(-i\zeta;1;-ips - i\vec{p}.\vec{s}), \tag{5}$$

with $\zeta = Z_p/p$ with $\vec{p} = \left|\vec{k} - \vec{v}\right|$ being the linear momentum of the ejected electron as seen from the projectile and where $\vec{s}$ denotes the position of the ejected electron with respect to the projectile nucleus[1].

Under these conditions, the doubly differential cross section of a given MO labeled *j*, namely, differential with respect to the ejected electron direction Ω and the emitted electron energy *E*, is given by

$$\sigma^{(2)}(\Omega, E) \equiv \left(\frac{d^2\sigma}{d\Omega dE}\right)_j = \frac{\mu k}{k_i}\int\left|T_{fi}\right|^2\delta\left(\frac{v^2}{2\mu} - \left|\varepsilon_\alpha^{(j)}\right| - \frac{k_s^2}{2\mu} - \frac{k^2}{2}\right)d\Omega_s\ , \tag{6}$$

where μ and k_s refer to the projectile mass and the momentum of the scattered particle, respectively. The scattering amplitude T_{fi} reported in equation (6) is given by

$$T_{fi} = \frac{-Z_P}{2\pi}\left\langle\chi_\beta^{-(i)}\middle|V\middle|\chi_\alpha^+\right\rangle \tag{7}$$

where *V* is the interaction between the incoming ion and the target.

Finally, for the multi-electronic target an independent active electron approximation is employed for describing the ionization process as it has been done in our previous work [14-15]. The passive target electrons (those non-ionized) are then considered frozen in their initial orbitals during the collision process. This assumption is generally made to overcome the difficulty of taking into account the dynamical correlation between active and passive electrons, in particular, for large molecules such as those investigated here (see also recent work of Monti *et al.* [21] where the influence of electron correlation on ionization of atomic targets has been analyzed in detail).

3.3. The target description and the cross section calculations

The input parameters used for describing the occupied molecular orbitals (MOs) of uracil were obtained by using *ab initio* methods (for more details we refer to our previous work [14-15]). Total-energy calculations were performed in the gas phase with the Gaussian 09 software at the RHF/3-21G level of theory [22]. The equilibrium geometries of the nucleobase were obtained without symmetry constraints applied while the computed ionization energies of the occupied molecular orbitals were scaled in a way that the calculated Koopmans ionization energy, *i.e.* the ionization

[1] *Note that the present equation (5) replaces the equation (2) of reference [15] where a typo has been recently detected in the argument of the second ${}_1F_1$ function.*

energy of the highest occupied molecular orbital (HOMO), coincides with the experimental value of the ionization potential measured by Hush and Cheung [23].

Thus, for each MO labeled *j*, the effective number of electrons $\xi_{j,i}$ relative to the atomic component *i* of the MO was derived from a standard Mulliken population analysis and their sum for each occupied MO is very close to 2, since only atomic shells with very small population have been discarded.

In the laboratory frame, the doubly differential cross sections can be written as

$$\sigma^{(2)}(\Omega,E)=\sum_{j=1}^{I=21}\sigma_j^{(2)}(\Omega,E), \tag{8}$$

where $\sigma_j^{(2)}$ refers to the doubly differential cross section of the *jth*-orbital, the latter being obtained as a weighted sum of atomic doubly differential cross sections corresponding to the different atomic components involved in the CNDO (complete neglected differential overlap) description, namely,

$$\sigma_j^{(2)}(\Omega,E)=\sum_i \xi_{j,i}.\sigma_{at,i}^{(2)}(\Omega,E). \tag{9}$$

Then, singly differential cross sections $\sigma^{(1)}(E)$ and total cross sections are obtained by successive numerical integrations over the ejected electron direction Ω and the emitted electron energy *E*. Note that in the present quantum mechanical calculations, an effective target charge Z_T^* is considered in the continuum factor describing the electron-residual target interaction, being $Z_T^*=\sqrt{-2n_\alpha^2\varepsilon_\alpha}$ where n_α refers to the principal quantum number of each atomic orbital component used in each MO expansion.

4. Results and discussions

In figure 2, we report absolute total ionization cross sections provided by the CB1-CWB and CDW-EIS descriptions, as a function of energy of the C-ions for two different charge states, namely, C^{4+} and C^{6+} [10]. We note that both sets of theories (solid and dashed lines) overestimate the experimental data even if the energy dependence is well reproduced. Besides, we also report in figure 2 the TCS provided by our previously published COB-CTMC model (dotted line). We note a better agreement than that reported by the quantum mechanical models in particular for bare C^{6+} projectiles. Indeed, the CTMC-COB model overestimates the experimental data only by a factor 1.2-1.4. Even for He-like C-ion this model gives a result closest to the experimental data, i.e. within a factor of ~1.5. The CDW-EIS and the CB1-CWB results differ from the experimental data by a factor of about 2 and 3, respectively. It is interesting to analyze the possible origin of this overestimation of the experimental data. In the case of proton beams impacting on uracil, CB1-CWB and CDW-EIS were shown to be in close agreement with measured total cross sections [15]. In principle, the effect could be attributed to the stronger projectile field created by a multi-charged ion in comparison with a proton beam. In CDW-EIS it is taken into account through the multiplicative factor distorting the initial bound orbitals, whereas in CB1-CWB this interaction is included only at asymptotic internuclear separations. For example, recently it has been shown for impact of multiply charged ions on He targets [24], that an excellent description of experimental data in the binary encounter region of double differential cross sections is obtained, if the eikonal phase of CDW-EIS is replaced by a corresponding continuum factor as in the CDW approximation. It is obvious that the models need to be fine-tuned in order to give a better agreement with the experiments.

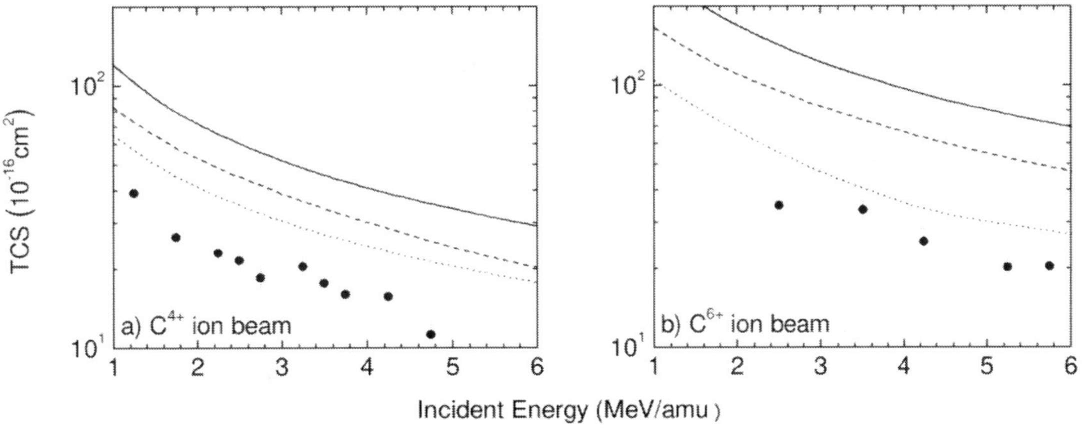

Figure 2. Comparison between the CB1-CWB, CDW-EIS and COB-CTMC TCS results (solid, dashed and dotted line, respectively) and the experimental measurements [10].

5. Conclusions

Single ionization of DNA-Uracil impacted by highly charged carbon ions has been studied by employing three different theoretical models, based on a CDW-EIS, CB1-CWB and CTMC-type approach, respectively.

A brief description of the experimental apparatus as well as the different theories has been reported while focusing our study on the differences between both quantum mechanical models.

An interesting experiment/theory confrontation has then clearly pointed out a theoretical overestimation of the experiment in particular when the CB1-CWB and the CDW-EIS models were used. Concurrently, the classical results have shown a better agreement with the experimental total ionization cross sections in particular when bare C^{6+} ions were studied.

This disagreement between theoretical predictions and measurements is a matter of our present interest and still remains an open question.

Acknowledgments

This work has been developed as part of the activities planned in the Programme de Coopération ECOS-*Sud* A09E04, the Project *Simulation Monte Carlo Haute Performance pour l'Hadronthérapie* partially funded by the Conseil Régional de Lorraine and the French Agence Nationale de la Recherche Contract No. ANR-09-BLAN-0135-01. Furthermore, some of the authors (MEG, OF and RDR) acknowledge partial support from the Agencia Nacional de Promoción Científica y Tecnológica (Project PICT 2006 No. 1912) and the Consejo Nacional de Investigaciones Cientificas y Técnicas (PIP-CONICET No. 1026/10 and No. 0033/10), both institutions from the República Argentina.

References

[1] Cucinotta F A, Wilson J W, Townsend L W, Shinn J L and Katz R 1992 *Nucl. Tracks Radial. Meas.* **20** 177-184
[2] Nikjoo H 2003 *Iran. J. Radiat. Res.* **1(1)** 3-16
[3] Friedland W, Dingfelder M, Kundrát P and Jacob P 2011 *Mutation Research* **711** 28-40
[4] Bernal M A, de Almeida C E, Sampaio C, Incerti S, Champion C and Nieminen P 2011 *Med. Phys.* **38(7)** 4147-4153
[5] Tabet J, Eden S, Feil S, Abdoul-Carime H, Farizon B, Farizon M, Ouaskit S and Märk T D 2010 *Phys. Rev.* A **81** 012711
[6] Tabet J, Eden S, Feil S, Abdoul-Carime H, Farizon B, Farizon M, Ouaskit S and Märk T D 2010 *Phys. Rev.* A **82** 022703

[7] Moretto-Capelle P and Le Padellec A 2006 *Phys. Rev.* A **74** 062705
[8] Iriki Y, Kikuchi Y, Imai M and Itoh A 2011 *Phys. Rev.* A **84** 032704
[9] Iriki Y, Kikuchi Y, Imai M and Itoh A 2011 *Phys. Rev.* A **84** 052719
[10] Agnihotri A N, Kasthurirangan S, Nandi S, Kumar A, Galassi M E, Rivarola R D, Fojón O, Champion C, Hanssen J, Lekadir H, Weck P F and Tribedi L C 2012 *Phys. Rev.* A **85** 032711
[11] Agnihotri A N, Kasthurirangan S, Nandi S, Kumar A, Galassi M E, Rivarola R D, Champion C, and Tribedi L C 2012 *Phys. Rev.* A (submitted)
[12] Abbas I, Champion C, Zarour B, Lasri B and Hanssen J 2008 *Phys. Med. Biol.* **53** N41
[13] Lekadir H, Abbas I, Champion C, Fojón O, Rivarola R D and Hanssen J 2009 *Phys. Rev.* A **79** 062710
[14] Champion C, Lekadir H, Galassi M E, Fojón O A, Rivarola R D and Hanssen J 2010 *Phys. Med. Biol.* **55** 6053-6067
[15] Galassi M E, Champion C, Weck P F, Fojón O A, Rivarola R D and Hanssen J 2012 *Phys. Med. Biol.* **57** 2081
[16] Agnihotri A N, Kelkar A H, Kasthurirangan S, Thulasiram K V, Desai C A, Fernandez W A and Tribedi L C 2011 *Phys. Scr.* **2011** 014038
[17] Bárány A, Astner G, Cederquist H, Danared H, Huldt S, Hvelplund P, Johnson A, Knudsen H, Liljeby L and Rensfelt K G 1985 *Nucl. Instr. and Meth. B* **9** 397
[18] Niehaus A 1986 *J. Phys. B* **19** 2925
[19] Abrines R and Percival I C 1966 *Proc. Phys. Soc.* **88** 861
[20] Stolterfoht N, DuBois R D and Rivarola R D 1997, *Electron emission in heavy ion-atom collisions*, Springer Series on Atoms and Plasmas, Ed. Springer-Verlag, ISSN 0177-6495
[21] Monti J M, Fojón O A, Hanssen J and Rivarola R D 2009 *J. Phys.* B **42** 195201
[22] Frisch M J *et al.* 2009 *Gaussian 09 Revision A.02 Gaussian Inc.* (Wallingford CT)
[23] Hush N S and Cheung A S 1975 *Chem. Phys. Lett.* **34** 11
[24] Monti J M, Rivarola R D and Fainstein P D 2011 *J. Phys.* B **44** 195206

Ion interaction with biomolecular systems and the effect of the environment

Alicja Domaracka, Michael Capron, Sylvain Maclot, Jean-Yves Chesnel, Alain Méry, Jean-Christophe Poully, Jimmy Rangama, Lamri Adoui, Patrick Rousseau and Bernd A. Huber

CIMAP, UMR 6252 (CEA/CNRS/ENSICAEN/Université de Caen Basse-Normandie), Bld Henri Becquerel, BP 5133, F-14070 Caen cedex 5, France

E-mail: domaracka@ganil.fr

Abstract. To fully understand the mechanisms of radiation damage in living tissues, a detailed knowledge of the processes occurring at the molecular level is needed. In the gas phase, most of the investigations concerning the ionization and fragmentation of biologically relevant molecular systems are performed with isolated molecules. The importance of such studies is limited to the intrinsic properties of these molecules because of the lack of a chemical environment. To probe the effect of such an environment on the behavior of small biomolecules under irradiation, the molecules (α-amino acids, adenine) were embedded into clusters. The present results, obtained with multiply charged ions, clearly indicate the protective role of the clusters since the total fragmentation yield is reduced for all systems. The surrounding molecules allow for a redistribution of the excess energy and of the charge within the cluster. In the case of adenine clusters, a new fragmentation channel is identified. Moreover, for hydrated adenine clusters, low-energy ion induced chemical reactions are observed, namely the proton transfer from the water cluster to the adenine molecule.

1. Introduction

To explore the molecular mechanisms underlying radiation damage, intensive studies concerning ionization and fragmentation of biomolecular systems have been performed in recent years. Such detailed knowledge is required in the context of cancer therapies (e.g. hadron-therapy) to investigate possible undesirable side effects and to possibly optimize the therapy strategies. A large fraction of radiation damage has been associated with effects of secondary particles (electrons, radicals, ions) which are created along the ion track.

It has been shown that slow electrons (with energy below the ionization threshold of the molecules) may lead to single or double DNA strand breaks [1,2]. Furthermore, small fragment ions originating from the break-up of nucleobases following collisions with highly charged ions were found to have a sufficient amount of kinetic energy (up to 100 eV) [3] to cause further harmful reactions within the DNA [4,5]. Therefore, ions with kinetic energies in the eV as well as in the keV range, are relevant for heavy ion induced biological radiation damage close to the region of the Bragg peak.

The interactions of different ionizing projectiles with isolated biomolecules have been addressed by means of gas-phase collision studies. This technique is excellent to obtain information about the fragmentation dynamics via mass spectrometric analysis. Unfortunately,

Published under licence by IOP Publishing Ltd

it neglects any effect of the chemical environment (presence of other biomolecules and surrounding water molecules) such as modified ionization energies [6, 7] and/or the dissipation of the excitation energy via solvent molecule evaporation (e.g. water). It has been shown that even a single water molecule can act as a catalyst in chemical reactions [8]. Therefore, in *real* biological systems the influence of the molecular environment has to be taken into account.

Embedding the molecule into a cluster of identical molecules or other molecules present in a realistic biological environment may lead one step further towards a more realistic case. Such an approach allows to keep the system under study simple enough to explore the underlying fundamental processes. For example, Liu *et al.* [9] have observed the protective effect of water molecules attached to the anion of RNA nucleotide adenosine 5'-monophosphate (AMP^-) when a such system undergoes Collision Induced Dissociation (CID). The energy transferred during the collision to the solvated molecule is used primarily to liberate the loosely bound water molecules, leading to a cooling of the residual system (the fragmentation yield of the AMP molecule was strongly reduced). Very surprisingly, this protective behavior was not at all observed in the case of Electron Capture Induced Dissociation (ECID) [10]. The AMP dianion is always fragmented (H-loss) regardless of the number of attached water molecules. This damaging effect becomes even more important as the number of initially attached water molecules increases [10].

The results for neutral clusters of nucleobases show that the presence of other molecules strongly influences the slow multiply charged ion induced fragmentation pathways. For example, in the case of thymine clusters two new dissociation channels were observed (loss of OH and NH_2CH) compared with the isolated thymine molecule. Such a behavior is interpreted as being caused by the weakening of intramolecular bonds due to the presence of an additional intermolecular interaction between the molecules in the cluster (hydrogen bonds) [11]. Therefore, by looking at these types of clusters environmental effects can be studied and the influence of different surroundings on molecular fragmentation can be clarified.

In this paper, we summarize environmental effects observed in collisions of different biomolecular clusters (glycine, α-alanine, valine, adenine) with slow multiply charged ions. Moreover, results for mixed adenine - water clusters will be compared with those obtained for pure adenine clusters as well as for the isolated molecule.

2. Experimental method

Experiments were performed at the low-energy ion beam facility ARIBE of GANIL (Caen, France). A brief overview of the experimental set-up is given here, for more details see reference [12].

The keV multiply charged ion beams, employed in our studies, were extracted from an electron cyclotron resonance (ECR) ion source. The mass selected, collimated and pulsed ion beam was guided to the collision zone where it crossed the neutral molecular beam.

Isolated molecular targets were produced by evaporation of a powder sample (Sigma-Aldrich, purity better than 98 %) in a heated oven. The temperature of the oven was chosen to assure sufficient vapor density and to avoid the thermal decomposition of the molecule. Alternatively, a modified gas phase aggregation source was used to produce amino acid and adenine clusters as well as hydrated adenine clusters. The biomolecules produced by an oven device entered a chamber filled with helium gas at the pressure of a few mbar. Molecules were then guided by the helium flow through a condensation channel (kept at liquid nitrogen temperature) where they aggregated. The neutral cluster beam was guided to the collision chamber by the helium flow through several differential pumping stages. To produce hydrated adenine clusters, water vapor was introduced to the helium gas injection line via a leak valve. To assure a sufficient water quantity a water tank was kept at 320 K and the injection line was heated to 340 K to avoid water condensation.

After the interaction, the cationic products were extracted into a modified Wiley-McLaren

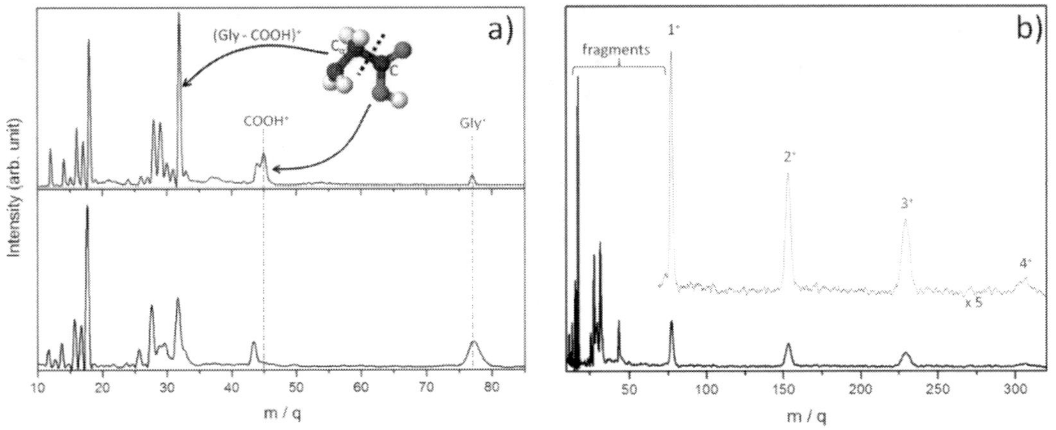

Figure 1. a) Comparison of the molecular fragmentation of the glycine molecule after the interaction with 300 keV Xe^{20+} ions: upper part – isolated molecule, lower part – cluster case. b) Cluster distribution shown in a wide-scale mass spectrum for collisions of 300 keV Xe^{20+} with neutral glycine clusters.

time-of-flight spectrometer [13], and were post-accelerated before impacting a gold-coated plate. The secondary electrons, produced at the plate, were deviated by a magnetic field towards a microchannel plate detector. This Daly-type detector assured a constant detection efficiency for cluster-ions with different masses and charges.

3. Results and discussion

3.1. Protective role of the cluster environment

The studies to identify the influence of the environment were performed in two steps: i) dissociation of the isolated molecule, ii) dissociation of the molecule embedded in a cluster. Figure 1b) displays a spectrum of the cationic cluster distribution (containing up to four molecules: Gly_n^+, n=1–4) obtained after collisions of 300 keV Xe^{20+} ions with neutral glycine clusters [14]. No multiply charged clusters or multiply charged fragments are observed. This finding will be discussed below. Furthermore, the absence of structures between cluster peaks indicates that there is no signal corresponding to clusters containing a fragment of the monomer. The observed production of small size fragments ($m<m_{monomer}$) probably occurs after the monomer has been emitted from the multiply charged cluster either in a fission or an evaporation process, similar to what has recently been reported for clusters of polycyclic aromatic hydrocarbons [15, 16].

In figure 1a) we compare the mass spectrum for masses below 85 amu for Xe^{20+} ion collisions with isolated glycine molecules and glycine clusters. In both cases we find the intact molecular glycine cation as well as small size fragments. In the cluster case, the width of the Gly^+ peak is larger compared to the molecular case (by a factor of 2), reflecting the kinetic energy acquired during the fragmentation process. The distribution of the small fragments is found to be different for the molecular and the cluster case. For example, the C-C_α bond cleavage, which leads to the formation of the $(Gly\text{-}COOH)^+$ cation (32 amu) and which is dominant in the molecular case (with a relative intensity of 18, see table 1), is strongly reduced for the cluster collision (relative intensity of 1.4 only). Also the complementary fragment $COOH^+$ is observed in the spectrum of the molecular target, but disappears totally in the cluster case. Thus, both observations suggest that the C-C_α bond of the glycine molecule is protected in the cluster case. The energy transferred during the collision is redistributed among the cluster constituents leading to the rupture of the hydrogen bonds between molecules (typical binding energy about hundreds of meV [17]) cooling the total system at the same time.

Table 1. Relative intensities of the main and the complementary fragments observed after collisions of low-energy multiply charged ions with glycine, α-alanine, valine and adenine as isolated molecules and as clusters, respectively. The intensities were normalized with respect to the monomer cation signal intensity. Data for isolated molecules and clusters of glycine and valine are from reference [14]. Results for isolated α-alanine molecules (20 keV He^{2+}) were taken from reference [18].

Target	Projectile	Isolated molecule	Cluster
Glycine	300 keV Xe^{20+}	$(Gly\text{-}COOH)^+$	$(Gly\text{-}COOH)^+$
		18	1.4
		$COOH^+$	$COOH^+$
		6	0
α-Alanine	32 keV He^{2+}	$(Ala\text{-}COOH)^+$	$(Ala\text{-}COOH)^+$
		∞	0.96
		$COOH^+$	$COOH^+$
		∞	0.09
Valine	300 keV Xe^{20+}	$(Val\text{-}COOH)^+$	$(Val\text{-}COOH)^+$
		∞	1
		$COOH^+$	$COOH^+$
		∞	0.05
Adenine	37.2 keV O^{3+}	$HNCH^+$	$HNCH^+$
		0.62	0.04
		$(Ade\text{-}HCNH)^+$	$(Ade\text{-}HCNH)^+$
		0.27	0.07

Furthermore, as to the fragments with masses below the monomer mass, the total fragmentation yield is also strongly reduced in the cluster case as shown in table 2. This decrease is not only due to the redistribution of the transferred energy, but also due to the charge repartitioning over the cluster volume. Recent ion-induced fragmentation studies of the amino acid molecules clearly show that the singly charged carboxyl group is mainly formed after multi-electron capture processes [19]. Thus, charge mobility explains why the $COOH^+$ fragment disappears in the cluster case. Multi-fragmentation processes (increasing markedly with the charge state of the glycine molecule) lead to the production of small fragments like C^+ or N^+ (the peaks in the range of m/q= 16–18 amu are due to collisions with the residual gas). Therefore, the strong decrease of the C^+ fragment intensity in the cluster case also indicates the high charge mobility within the cluster.

As indicated in tables 1 and 2, we have also performed experiments with other biomolecular systems: the α-amino acids (α-alanine ($NH_2CH_2CH_3COOH$) and valine ($NH_2(CH)_2(CH_3)_2COOH$)), and the nucleobase adenine ($C_5H_5N_5$). The findings are similar to those observed for glycine. It should be noted that for isolated α-alanine and valine molecules, the cationic intact molecule (M^+) is not observed [14,18]. Similar results have been obtained with other ionizing methods [20–23]. The main dissociation channel occurs via C-C_α bond cleavage, where the loss of neutral COOH takes place (formation of the decarboxylated ion, $(M\text{-}COOH)^+$). The $COOH^+$ cation is also observed with lower intensity. As α-alanine and valine molecules do not survive the collision, the intensity ratios presented in the table 1 for $(M\text{-}COOH)^+$ and $COOH^+$ channels are equal to ∞ and total fragmentation yields are 1, respectively.

Table 2. Relative intensities of singly charged carbon ions (I_{C^+}/I_T) and of the total fragmentation yield (I_F/I_T) for isolated biomolecules and molecules embedded in a cluster environment. I_T is the integral intensity of the measured spectrum for masses above 10 amu. The projectiles correspond to those given in table 1.

Target	I_{C^+}/I_T		I_F/I_T	
	Isolated molecule	Cluster	Isolated molecule	Cluster
Glycine	0.07	0.04	0.98	0.84
α-Alanine	0.04	0.01	1	0.61
Valine	0.06	0.01	1	0.77
Adenine	0.1	0.01	0.79	0.39

In summary, the present results cleary indicate *the protective role of the cluster environment.* The total fragmentation yield is significantly reduced (up to 50%). Moreover, embedding the α-alanine and valine molecule into clusters allows to protect the C-C_α bonds, and intact molecular cations are observed. In the same way as for glycine clusters, the $(M\text{-}COOH)^+$ signal intensities are reduced and the quenching of the $COOH^+$ fragments is observed.

Figure 2 shows the mass spectra of the cations produced in the interaction of O^{3+} ions with isolated adenine (upper part) and neutral adenine clusters (lower part). The choice of the region of interest (from $m/q = 10$ to 140 amu) facilitates the comparison of the fragmentation patterns. In both cases, fragments are divided into series containing a fixed number of heavy atoms (carbon and/or nitrogen - the number is indicated on the top of figure 2) and a varying number of hydrogen atoms. The fragmentation of isolated adenine molecules is mainly driven by the successive loss of neutral HCN fragments [24–27]. The fragmentation of adenine within the cluster follows the same pattern. However, from a global point of view, the fragmentation is considerably reduced by a factor of 2 (table 2 and figure 2), showing the protective effect already presented for amino acids. For example, the main fragmentation channel – formation of $HCNH^+$ (28 amu) – has been lowered by a factor of 15 (table 1). As mentioned before, the cluster dissipates the excess energy by evaporation of monomers and/or by cluster fission where the hydrogen bonds break preferably. The intensity ratios of carbon ions to the total yield of ionic products (I_{C^+}/I_T), like in the case of amino acid clusters, are substantially reduced indicating that a rapid charge redistribution takes place within the clusters. This statement is also supported by the fact that doubly charged adenine ions (Ade^{2+}) are not observed any more for the cluster case.

3.2. New fragmentation channels due to cluster environment

Nevertheless, protection is not the only environmental effect that appears in the case of adenine clusters. The intensity of the fragment signals containing 9 and 7 heavy atoms are indeed stronger in the cluster case (see figure 2). Such a behavior for nucleobase clusters has already

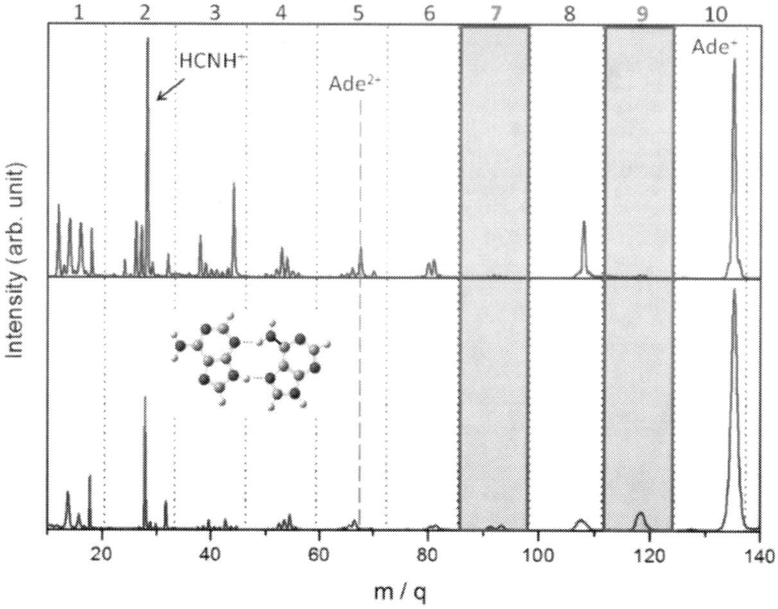

Figure 2. Cationic products of the interaction between 37.2 keV O^{3+} ions and isolated adenine molecules (upper part) and adenine clusters (lower part). The numbers given on the top of the graph correspond to the number of heavy atoms in a fragment. The insert displays a neutral adenine dimer which has been observed experimentally [28]. Regions in red indicate new fragmentation channels observed for adenine molecules embedded in the cluster.

been mentioned by Schlathölter *et al.* [11]. The opening of additional fragmentation channels for the adenine molecule is related to the interplay between *intra-* and *inter*molecular interactions. It is shown that in hydrogen bonded regions, the electronic structure of the molecule is strongly modified [29]. The IR-UV spectrum of the adenine dimer indicates a molecular structure where molecules are bound by a N···HN and a C-NH_2···N bond [28]. Such an interaction weakens the neighbored C-NH_2 bond and gives rise to the fragment ions $(\text{Ade-NH}_n)^+$ with masses of 118 ($n = 3$) and 119 amu ($n = 2$). In a second step, this fragment may lose an HCN unit leading to $C_4H_2N_3$ where an additional H atom might be lost or added (91 and 93 amu). Similar results have been obtained for uracil and thymine molecules [11].

3.3. Proton transfer within hydrated adenine clusters

In general, biological reactions occur in aqueous solutions. Therefore, hydration plays a very important role in the chemistry of biological systems. A better understanding of radiation damage depends on the knowledge of the physical and chemical behavior of the hydrated complexes consisting for example of nucleobases and water molecules. Over the last years, intensive efforts were made to study the water binding sites and energies, the effect of water molecules on conformation of biomolecules, as well as the effect on their stability (e.g. [9, 10, 30–33]).

Here, we present preliminary results for low-energy ion interaction (37.2 keV O^{3+}) with neutral mixed adenine-water clusters. To our knowledge, these are the first results on nucleobase-water complexe dissociation in the gas phase induced by low-energy ions. Figure 3 displays a part of the observed mass spectrum, showing the mass range between the adenine monomer and the

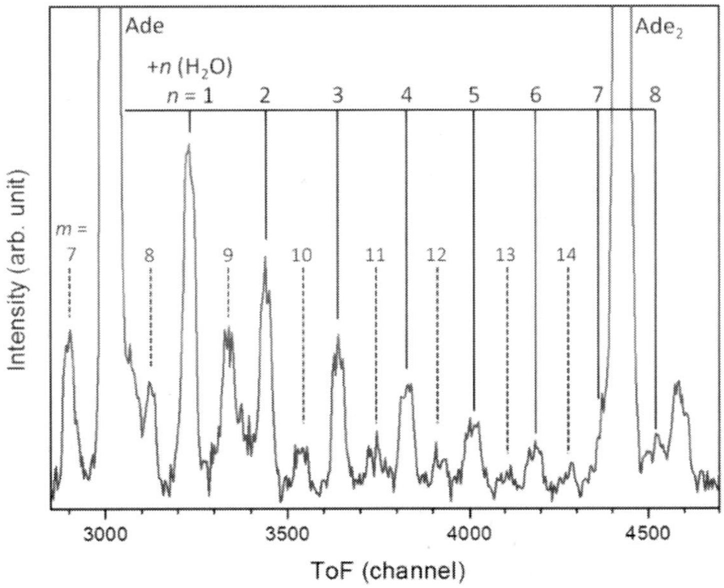

Figure 3. Cationic products of the interaction between 37.2 keV O^{3+} ions and neutral hydrated adenine clusters shown in the range between the adenine monomer (Ade) and dimer (Ade_2). For more details see text.

adenine dimer. In addition to these two peaks, we observe two series of peaks which correspond to protonated water clusters ($H^+(H_2O)_m$) with m=7 to 14 and the adenine monomer to which up to n=8 water molecules are attached ($Ade(H_2O)_n$). In general, these clusters are fragments from larger clusters which evaporate or fragment due to the energy and charge transferred in the collision. This is partly indicated by the increased peakwidth of the adenine peak compared to ion collisions with an isolated molecule. The fact that during these fragmentation processes water molecules remain linked to the adenine, at a microsecond time scale, is not surprising. Indeed, binding energies between water molecules, adenine molecules and between water and adenine molecules within clusters have similar values [34, 35]. Compared to the non-hydrated cluster case, the fragmentation yield (fragments with masses below the adenine monomer, not shown in figure 3) is lower and these low-mass fragments are not observed with a water molecule attached. This suggest that the fragmentation of the adenine molecule occurs after liberation of all loosely bound water molecules. Moreover, this evaporation chain leads to a cooling of the system. Therefore, the fragmentation probability of the adenine molecule itself is lowered.

We should mention that similar mass spectra have been obtained by electron [36, 37] and by photon impact [34,37]. In the latter case it was shown that the intensity of the hydrated adenine monomer depends strongly on the laser wavelength.

In figure 4, we compare the peak width of the adenine monomer for spectra obtained with isolated molecules, pure adenine clusters and hydrated adenine clusters. As already mentioned above, for the isolated molecule the peak is rather narrow showing its isotopic contributions and the loss of a hydrogen atom. For pure adenine clusters the monomer peak is widened due to evaporation/fragmentation processes preventing the observation of any fine structure. In the case of hydrated adenine clusters the peak is further widened and its center is shifted by about half a mass unit. This finding is interpreted in the following way: the monomer adenine peak of the hydrated cluster spectrum is in fact the result of two different contributions – the adenine molecule (Ade^+) as well as the protonated adenine ion ($AdeH^+$). As we do not observe proto-

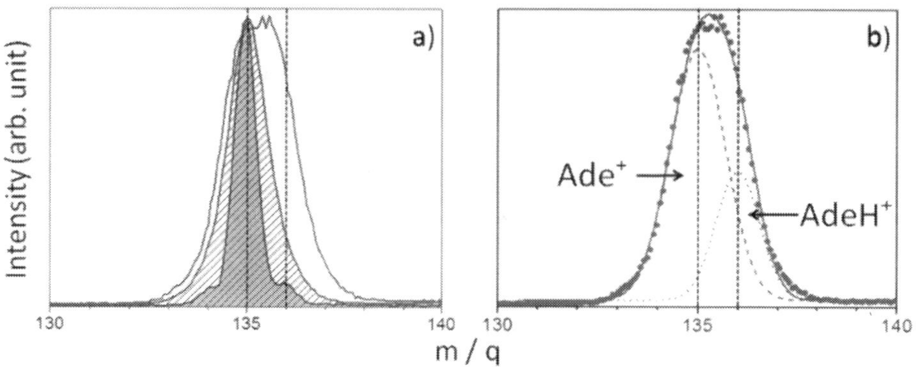

Figure 4. Zoom into the m/q = 130–140 amu mass range, showing the adenine monomer after the collision with 37.2 keV O^{3+} ions. a) Grey area - isolated molecule, hatched area - adenine clusters, and white area - hydrated adenine clusters. b) Large blue dots correspond to experimental points of the adenine monomer for hydrated adenine clusters. Full blue line represents the sum of fitting obtained for the contribution of adenine ion (Ade^+, m/q = 135 amu) and protonated adenine ion ($AdeH^+$, 136 amu).

nation in the pure cluster case (see figure 4a)), we assume that protonation of adenine (b) is due to the presence of water molecules. The protonated adenine monomer peak was also observed in the case of electron impact and multiphoton ionization [34, 37, 38]. The electron impact studies of hydrated nucleobases suggest that the protonated nucleobase monomer is only observed when the electron energy is above the ionization potential of water [36]. Therefore, to explain such a formation, we propose the following mechanism in the case of single electron capture for a water molecule within an hydrated adenine cluster:

$$O^{3+} + [Ade + (H_2O)_n] \longrightarrow O^{2+} + [Ade + (H_2O)^+_n]^* \quad (1)$$

$$[Ade + (H_2O)^+_n]^* \longrightarrow [Ade + H^+(H_2O)_{n-1}]^* + OH^* \quad (2)$$

$$\searrow \qquad \downarrow$$

$$[AdeH^+ + (H_2O)_{n-1}]^* + OH^* \quad (3)$$

$$[AdeH^+ + (H_2O)_{n-1}]^* \longrightarrow [AdeH^+]^* + [(H_2O)_{n-1}]^* \quad (4)$$

First, the O^{3+} ion interacts with an hydrated cluster of adenine, removing an electron from one of the water molecules (equation (1)). Then, a protonated cluster is fastly formed by evaporation of a neutral OH fragment at the picosecond time scale (such a behavior was observed in water clusters [39]). As the proton affinity of the adenine molecule is higher than that of water [40, 41], a proton transfer to the adenine molecule may occur (see equation (2) and (3)). The protonated adenine monomer ($AdeH^+$) is observed after losing all the surrounding water molecules (4). Experiments performed with 300 keV Xe^{20+} ions have shown similar results. The contribution of protonated adenine compared to Ade^+ was increased. It indicates that the number of electrons captured from the water molecules strongly influences this process.

4. Conclusions

We have studied the fragmentation of α-amino acids and adenine molecules in the collision with low-energy multiply charged ions. We evidenced that the presence of an environment strongly influences the fragmentation dynamics of the molecules. On the one hand, the energy transferred during collisions is dissipated by evaporation and/or cluster fission leading to a protection of the C-C_α bond in the molecule. The hydrogen bonds in the cluster are preferentially broken. On the second hand, the fast charge redistribution within the cluster leads to a protection of the molecule as well (for example the $COOH^+$ channel is quenched for amino acids).

For clusters of adenine molecules we have observed new fragmentation channels related to the weakening of the intermolecular bond due to hydrogen bonding of the clusters. Furthermore, we have observed ion-induced chemical reactions within hydrated adenine clusters, leading to a protonation of the adenine monomer.

Acknowledgments

The experimental studies have been performed at the ARIBE facility, the low-energy ion facility of GANIL (France). Financial support received from the ITS-LEIF project (Grant No. RII3-026015), the ANR Programme Blanc PIBALE (ANR-09-BLAN-0130-01) and the Conseil Régional de Basse-Normandie is gratefully acknowledged.

References

[1] Boudaïffa B, Cloutier P, Hunting D, Huels M and Sanche L 2000 *Science* **287** 1658
[2] Huels M, Boudaïffa B, Cloutier P, Hunting D and Sanche L 2003 *J. Am. Chem. Soc.* **125** 4467
[3] Schlathölter T, Hoekstra R, Zamith S, Ni Y, Muller H and Vrakking M 2005 *Phys. Rev. Lett.* **94** 233001
[4] Deng Z, Imhoff M and Huels M 2005 *J. Chem. Phys.* **123** 144509
[5] Deng Z, Bald I, Illenberger E and Huels M 2005 *Phys. Rev. Lett.* **95** 153201
[6] Belau L, Wilson K, Leone S and Ahmed M 2007 *J. Phys. Chem.* A **111** 7562
[7] Pluhařová E, Jungwirth P, Bradforth S and Slavíček P 2011 *J. Chem. Phys.* **115** 1294
[8] Vöhringer-Martinez E, Hansmann B, Hernandez H, Francisco J, Troe J and Abel B 2007 *Science* **317** 497
[9] Liu B, Brondsted-Nielsen S, Hvelplund P, Zettergren H, Cederquist H, Manil B and Huber B 2006 *Phys. Rev. Lett.* **97** 133401
[10] Liu B, Haag N, Johansson H, Schmidt H T, Cederquist H, Brøndsted-Nielsen S, Zettergren H, Hvelplund P, Manil B and Huber B A 2008 *J. Chem. Phys.* **128** 075102
[11] Schlathölter T, Alvarado F, Bari S, Lecointre A, Hoekstra R, Bernigaud V, Manil B, Rangama J and Huber B 2006 *ChemPhysChem* **7** 2339
[12] Bergen T, Biquard X, Brenac A, Chandezon F, Huber B, Jalabert D, Lebius H, Manurel M, Monnand E, Opitz J, Pesnelle A, Pras B, Ristori C and Rocco J 1999 *Rev. Sci. Instrum.* **70** 3244
[13] Chandezon F, Huber B and Ristori C 1994 *Rev. Sci. Instrum.* **65** 3344
[14] Maclot S, Capron M, Maisonny R, Lawicki A, Méry A, Rangama J, Chesnel J Y, Bari S, Hoekstra R, Schlathölter T, Manil B, Rousseau P and Huber B 2011 *ChemPhysChem* **12** 930
[15] Holm A, Zettergren H, Johansson H, Seitz F, Rosén S, Schmidt H, Lawicki A, Rangama J, Rousseau P, Capron M, Maisonny R, Adoui L, Méry A, Manil B, Huber B and Cederquist H 2010 *Phys. Rev. Lett.* **105** 213401
[16] Johansson H A B, Zettergren H, Holm A I S, Seitz F, Schmidt H T, Cederquist H, Rousseau P, Lawicki A, Capron M, Domaracka A, Lattouf E, Maclot S, Maisonny R, Manil B, Chesnel J Y, Adoui L, Huber B A and Cederquist H 2011 *Phys. Rev.* A **84** 043201
[17] Chocholoušová J, Vacek J, Huisken F, Werhahn O and Hobza P 2002 *J. Phys. Chem.* **106** 11540
[18] Bari S, Sobocinski P, Postma J, Alvarado F, Hoekstra R, Bernigaud V, Manil B, Rangama J, Huber B and Schlathölter T 2008 *J. Chem. Phys.* **128** 074306
[19] Capron M, Díaz-Tendero S, Maclot S, Domaracka A, Lattouf E, Lawicki A, Maisonny R, Chesnel J Y, Méry A, Poully J C, Rangama J, Adoui L, Martın F, Alcami M, Rousseau P and Huber B A 2012 *Chem. Eur. J.* accepted (DOI: 10.1002/chem.201103922)
[20] Junk G and Svec H 1963 *J. Am. Chem. Soc.* **85** 839
[21] *NIST Chemistry WebBook (Eds.: P.J. Linstrom, W.G. Mallard), NIST Standard Reference Database Number 69, Institute of Standards and Techniology, http://webbook.nist.gov*
[22] Vorsa V, Kono T, Willey K and Winograd N 1999 *J. Phys. Chem.* B **103** 7889

[23] Jochims H W, Schwell M, Chotin J L, Clemino M, Dulieu F, Baumgärtel H and Leach S 2004 *Chem. Phys.* **298** 279
[24] Rice J and Dudek G 1967 *J. Am. Chem. Soc.* **89** 2719
[25] Jochims H W, Schwell M, Baumgärtel H and Leach S 2005 *Chem. Phys.* **314** 263
[26] Alvarado F, Bari S, Hoekstra R and Schlathölter T 2007 *J. Chem. Phys.* **127** 034301
[27] Martin S, Brédy R, Allouche A, Bernard J, Salmoun A, Li B and Chen L 2008 *Phys. Rev.* A **77** 062513
[28] Plützer C, Hünig I and Kleinermanns K 2003 *Phys. Chem. Chem. Phys.* **5** 1158
[29] Kelly R, Lee Y and Katorovich L 2005 *J. Phys. Chem.* B **109** 11933
[30] Sukhodub L 1987 *Chem. Rev.* **87** 589
[31] Schneider B and Berman H 1995 *Biophys. J.* **69** 2661
[32] Prell J S, O'Brien J T, Holm A I S, Leib R D, Donald W A and Williams E R 2008 *J. Am. Chem. Soc.* **130** 12680
[33] Wyttenbach T and Bowers M 2009 *Chem. Phys. Lett.* **480** 1
[34] Nam S H, Park H S, Ryu S, Song J K and Park S M 2008 *Chem. Phys. Lett* **450** 236
[35] Preuss M and Bechstedt F 2008 *Surf. Sci.* **602** 1643
[36] Kim S K, Lee W and Herschbach D R 1996 *J. Phys. Chem.* **100** 7933
[37] Kim N, Kang H, Jeong G, Kim Y S, Lee K T and Kim S K 2000 *J. Phys. Chem.* A **104** 6552
[38] Nam S H, Park S S, Song J K and Park S M 2007 *J. Phys. Chem.* A **111** 3480
[39] Adoui L, Cassimi A, Gervais B, Grandin J P, Guillaume L, Maisonny R, Legendre S, Tarisien M, López-Tarifa P, Politis M F, du Penhoat M A H, Vuilleumier R, Gaigeot M P, Tavernelli I, Alcamí M and Martín F 2009 *J. Phys.* B **42** 075101
[40] Meot-Ner M 1979 *J. Am. Chem. Soc.* **101** 2396
[41] Chang H C, Jiang J C, Hahndorf I, Lin S H, Lee Y T and Chang H C 1999 *J. Am. Chem. Soc.* **121** 4443

Nano-Sensitization under gamma rays and fast ion radiation

E PORCEL[1], S LI[1], N USAMI[2], H REMITA[3], Y FURUSAWA[4], K KOBAYASHI[2], C LE SECH[1] and S LACOMBE[1]

[1] Institut des Sciences Moléculaires d'Orsay (UMR 8214) Bât 351, Université Paris Sud 11, CNRS, 91405, Orsay Cedex, France.
[2] Photon Factory, Institute of Materials Structure Science, High Energy Accelerator Research Organization, Oho, Tsukuba, Ibaraki 305-0801, Japan.
[3] Laboratoire de Chimie Physique (UMR 8000) Bât 349, Université Paris Sud 11, CNRS, 91405, Orsay Cedex, France.
[4] Research Center for Charged Particle Therapy, National Institute of Radiological Sciences, 4-9-1 Anagawa, Inage-ku, Chiba 263-8555, Japan.

E-mail: sandrine.lacombe@u-psud.fr

Abstract. The use of heavy compounds to enhance radiation induced damage is a promising approach to improve the therapeutic index of radiotherapy. In order to quantify and control the effects of these radiosensitizers, it is of fundamental interest to describe the elementary processes which take place at the molecular level. Using DNA as a probe, we present a comparison of the damage induced in the presence of platinum compounds exposed to different types of ionizing radiation. We present the results obtained with gamma rays (Linear Energy Transfer (LET) = 0.2 keV.μm^{-1}), fast helium ions He^{2+} (LET = 2.3 keV.μm^{-1}) and fast carbon ions C^{6+} (LET =13 keV.μm^{-1} and LET=110 keV.μm^{-1}). The efficiency of two different sensitizers was measured: platinum based molecules (the chloroterpyridine platinum - PtTC) and platinum nanoparticles (PtNP). These experiments show that the two sensitizers are efficiently amplifying molecular damage under photon or ion irradiation. Experiments with a radical scavenger confirmed that these damages are mediated by free radicals for more than 90%. More interestingly, the induction of complex damage, the most lethal for the cells, is amplified by a factor of 1.5 on average if platinum (PtTC and PtNP) is present. As already known, the induction of complex damages increases also with the radiation LET. So, finally, the most significant enhancement of complex damage is observed when ion radiation is combined with platinum induced sensitization.

1. Introduction

Ionizing particles deliver their energy along their tracks and induce different inelastic processes like excitations or ionizations of atoms and molecules in the irradiated material. These inelastic mechanisms result in the emission of energetic secondary electrons which can induce subsequent breaks of chemical bounds. In tissues of living organisms cell death might result as a consequence of the irradiation and this property is used for therapeutic purposes to treat tumors by radiotherapy.

In order to improve the efficiency of radiotherapy one possible way is to load the tumoral tissues with high-Z atoms. Molecules containing high-Z atoms, as for example gold salts, are used for medical purposes. Chemotherapy protocols frequently use platinum salts such as cis-platinum. Our group has previously shown that platinum salts enhance DNA damage and cell death induced by radiation (X-rays or fast ions) [1]. An inner shell ionization of the heavy atom by the ionizing particles or by the secondary electrons emitted along the track will mainly generate an Auger cascade and result in the amplification of electron emission. Subsequently, it will induce an enhancement of the water radiolysis in the vicinity of the heavy atom [2]. The local release of a dense cluster of water radicals is very efficient to cause complex damage in the surrounding molecules. This promising strategy is however limited by the localization of the sensitizers in the tumor. Recently, some developments in nanoscience opened new perspectives by using nanoparticles, whose surfaces can be functionalized by tumor specific molecules [3, 4]. Nanoparticles made of heavy atoms are also of interest to achieve the radiosensitization of biological tissues. They are presently the subject of an increasing number of studies [5-8]. Most of these studies are focused on the effects of X-rays. However, such highly penetrating photons induce severe damage before (in the entrance channel) and behind the tumors. An alternative and superior approach consists of using fast ions as ionizing radiation [9].

In this work, we compare the efficiency of sensitization induced by platinum salts and platinum nanoparticles when gamma rays (LET = 0.2 $keV.\mu m^{-1}$) or fast helium and carbon ions (LET = 2.3, 13 and 110 $keV.\mu m^{-1}$) are used as ionizing radiation sources. Studies at the molecular level shall help to understand the elementary mechanisms. Recent microscopy experiments have clearly shown that the radiosensitization is not limited to the compartment of nuclear DNA [10, 11]. We use, however, DNA as a good probe to quantify simultaneously the induction of simple and complex damage. In particular, the induction of complex (nanometric) damage is of high interest because they induce cell death. The goal of this work is thus to quantify the molecular damage induced by combining radiation of different LET and platinum induced radiosensitization.

2. Experimental section

2.1. DNA samples
Plasmid DNA pBR322 (supplied by Toyobo or Promega) is extracted from E. Coli. They are supplied in Tris EDTA (TE) buffer (10 $mmol.L^{-1}$ Tris-HCl –pH 7.6– and 1 $mmol.L^{-1}$ ethylenediaminetetraacetic acid –EDTA–). This buffer is chosen because of the presence of chlorine, which mimics the medium in living cells. The molecule is a circular double-stranded DNA of 4361 base pairs (2.8×10^6 Dalton per molecule). Three forms of plasmid can be found depending on the degradation level. In the intact form of the supercoiled plasmid one strand has more turns as compared to the other. This results in a topological constraint that can be removed only by strand breaks. A single strand break (SSB) relaxes the constraint and the plasmid becomes "circular". A double strand break DNA (DSB) consists of two breaks occurring face to face - within about ten base pairs - in the two opposite strands separated by 2 nm and the plasmid becomes open in a form called "linear". The three forms can be easily separated by electrophoresis performing a migration in agarose gel. Prior to irradiation, DNA samples contain more than 95% supercoiled, 5% circular, and no linear forms.

2.2. Chloroterpiridine platinum
This molecule is provided by FLUKA –Sigma Aldrich Chemie Gmbh Schnelldorf Germany– in powder form and used without any further purification. A solution of PtTC was prepared by dilution in ultra pure water. The concentration was determined by spectrophotometry at the wave length $\lambda = 278$ nm with an absorbance coefficient $\varepsilon = 25100\ mol^{-1}.L.cm^{-1}$ [12]. Dilutions were made to obtain a final solution of a concentration of $4.23\times 10^{-5}\ mol.L^{-1}$.

2.3. Platinum nanoparticles
Platinum nanoparticles (PtNP) were synthesized by radiolysis [13]. The synthesis is described elsewhere [14]. Briefly, the PtNP were made from platinum salts $[Pt(NH_3)_4]Cl_2$, H_2O –Sigma Aldrich– diluted in ultra pure water. A polymer, the poly acrylic acid (PAA) $C_3H_4O_2C_3H_4O_2Na$ –Sigma Aldrich–, was added to stop aggregation and stabilize the aggregate. The solution prepared at a

platinum concentration of 10^{-4} mol.L^{-1} was irradiated by gamma rays to induce the production of reducing species (e-, H^{+}). The dose needed for complete reduction of platinum (+II) in platinum (0) was 1000 Gy. After aggregation of platinum (0), homogeneous nanoparticles of 3 nm in diameter and containing 1000 platinum atoms on average were obtained [14].

2.4. Preparation of the samples
Solutions of DNA molecules containing platinum were prepared with a ratio of one platinum atom per 15 DNA phosphate groups (one platinum atom for 7 or 8 base pairs on average). The experiments with PtTC and PtNP were performed with the same atomic concentration of platinum in DNA.

An electrostatic binding is established between DNA (negative) and PtTC (positive) [12]. At this concentration, a negligible fraction of PtTC remains free in solution and a complete saturation of DNA by PtTC is observed [15]. A similar type of electrostatic binding is expected for PtNP.

In order to investigate the role of the indirect effects (mediated by the solvent radicals), some experiments were performed in the presence of the free radical scavenger, Dimethylsulfoxide (DMSO), at a concentration of 1 mol.L^{-1}.

Finally, the samples containing 500 ng of DNA (1 µL) were diluted in 12 µL of TE buffer at room temperature. The PtTC and PtNP solutions were added to DNA at a same final concentration of 5.6×10^{-6} mol.L^{-1} (1 platinum for 15 DNA phosphate groups). In order to keep the concentration in salts (TE buffer) similar in all the experiments, the total volume of the samples was adjusted to 18 µL with pure water (used as the solvent of the sensitizers and DMSO).

2.5. Irradiation
The DNA solutions were placed in Eppendorf vessels of 0.5 mL. Irradiations of these samples were respectively performed with gamma rays, and fast ions (He^{2+} and C^{6+}). The experiments were performed with pure DNA, DNA loaded with PtTC and DNA loaded with PtNP. The irradiations were conducted under atmospheric conditions at room temperature. Samples were irradiated with increasing doses ranging from 0 to 400 Gy.

The gamma irradiation of DNA was performed in a panoramic ^{60}Co source located at Paris Sud University (Orsay - France). The average energy of the photon beam is 1.25 MeV and the LET is 0.2 keV.µm^{-1}. Samples were irradiated at the maximum dose rate (33 Gy.min^{-1}).

The irradiation of DNA by fast ions were performed at the Heavy Ion Medical Accelerator (HIMAC) located in Chiba, Japan. DNA was irradiated by ions of different Linear Energy Transfer (LET): He^{2+} at the entrance channel of the beam (Energy = 150 MeV/uma, LET = 2.3 keV.µm^{-1}), C^{6+} at the entrance of the beam (Energy = 290 MeV/uma, LET = 13 keV.µm^{-1}) and C^{6+} at the maximum of the Spread Out Bragg Peak – SOBP – (Initial energy = 290 MeV/uma, average LET = 110 keV.µm^{-1}). The dose rates were equal to 4 Gy.min^{-1} for helium and 15 Gy.min^{-1} for carbon.

2.6. Analysis
After irradiation, the samples were analyzed by agarose gel electrophoresis. The gel consisted of 1% agarose in a buffer: Tris acetate EDTA (40 mmol.L^{-1} Tris, 20 mmol.L^{-1} acetate and 1 mmol.L^{-1} EDTA). After 2h30 of migration at room temperature (10 V.cm^{-1}), the gel is stained with Ethidium Bromide at 0.3 µg.mL^{-1} and placed under Ultra Violet light (312 nm). The DNA fluorescence was recorded with a Charged Coupled Device (CCD) camera. An image analysis software (Image Quant) was used to determine the integrated fluorescence of the three bands corresponding to the three plasmid conformations. Considering Ethidium bromide binds 1.47 times less on supercoiled plasmids *–S–* than in relaxed *–R–* and linear *–L–* forms, the normalized fractions of supercoiled plasmids *–S'–* relaxed *–R'–* and linear *–L'–* forms were obtained and the normalized fractions of single (*SSBpl*) and double strand breaks (*DSBpl*) per plasmid were finally calculated according to [16]:

$$\text{Norme : } N = S\times 1.47 + R + L$$

$$S' = \frac{S}{N} \qquad R' = \frac{R}{N} \qquad L' = \frac{L}{N}$$

$$SSBpl = \ln\left(\frac{1-L'}{S'}\right) \qquad DSBpl = \frac{L'}{1-L'}$$

The final outcome is the number of SSB and DSB per plasmid versus the dose. No significant artefacts due to the binding of PtTC or PtNP to DNA were found in electrophoresis. Furthermore, the different radiosensitizers did not cleave the plasmids without irradiation.

3. Results and discussion

3.1. Amplification of radiation effects

The radiosensitizing properties of platinum compounds were characterized by measuring the induction of the DNA breaks: single strand breaks (SSB) and double strand breaks (DSB). DSB is the signature of nanometer scale damage, difficult to repair by the cells, while SSB corresponds to a simple bond breaking. For this purpose, plasmid DNA is used as a sensitive probe to quantify and to characterize simultaneously this damage induced by radiation, with and without sensitizers.

The number of SSB and DSB in DNA plasmids loaded with or without platinum nanoparticles or platinum atoms induced by gamma rays (LET = 0.2 keV.μm^{-1}), by helium ions He^{2+} (LET = 2.3 keV.μm^{-1}), by carbon ions C^{6+} (LET = 13 keV.μm^{-1}), and by carbon ions C^{6+} (LET = 110 keV.μm^{-1}) have been measured as a function of the dose. Some results, obtained with gamma rays and helium ions are presented in figure 1. The calculations related to these measurements are summarized for all types of irradiation in table 1.

For every radiation condition, the number of breaks (SSB, DSB) increases with the dose and follows a linear law. This finding suggests that a single event is able to induce a double strand break. In the presence of platinum (salts and nanoparticles), an enhancement of DNA damage is observed. The measurement is in agreement with previous work [14, 17].

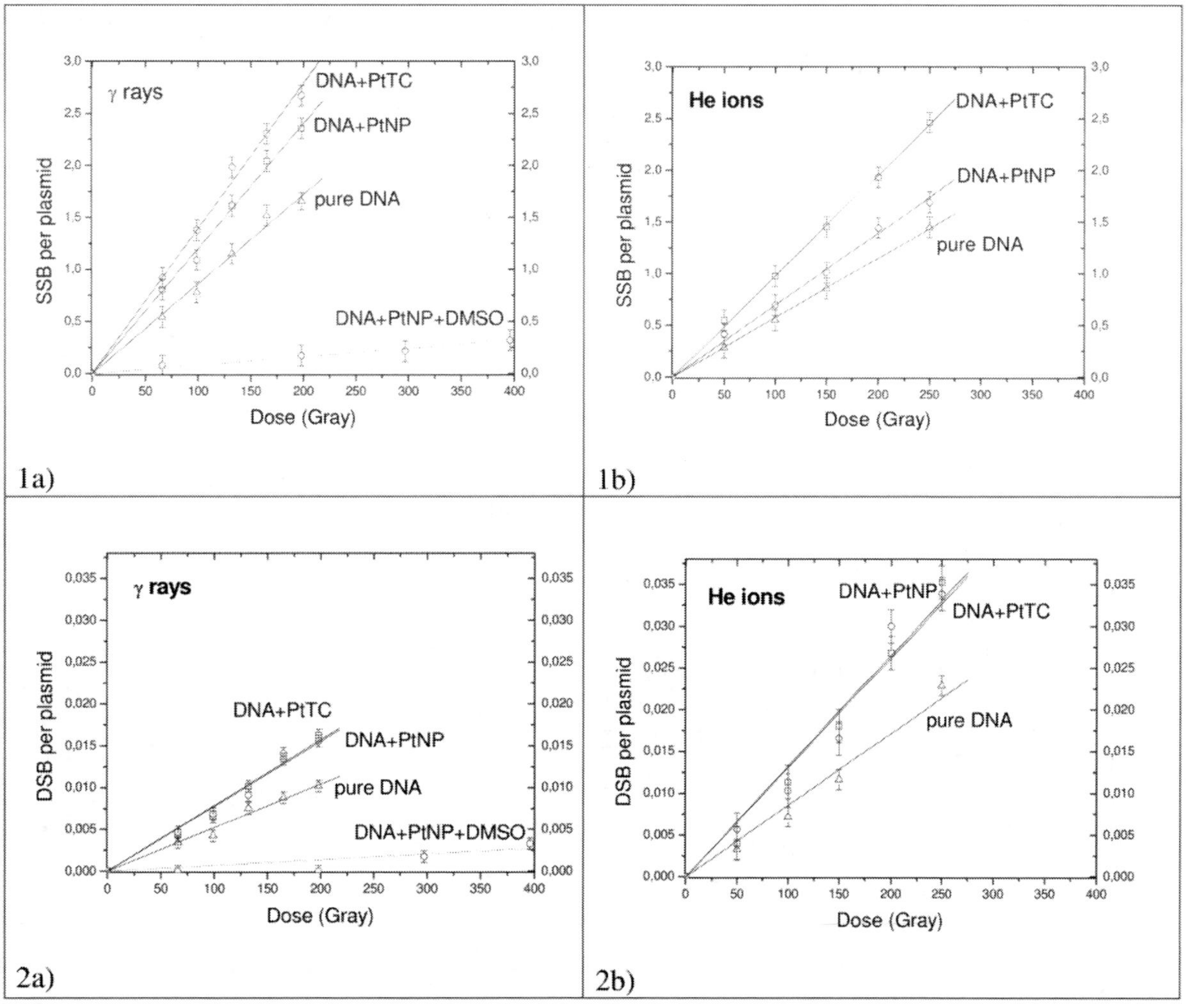

Figure 1. SSB (1a and 1b) and DSB (2a and 2b) per plasmid induced by gamma rays (energy = 1.25 MeV, LET = 0.2 keV.μm^{-1}) irradiation (a: left column) and He^{2+} (energy = 150 MeV/uma, LET = 2.3 keV.μm^{-1}) (b: right column) in the pure DNA (open triangle), in the presence of PtNP (open diamond) or PtNP + DMSO (open circles) and in the presence of PtTC (open square).

The yields of SSB and DSB induced to DNA are defined as the number of breaks induced per plasmid and per gray. They correspond to the slopes of the dose–response curves as presented in figure 1 for gamma rays and helium ions. The yields are reported in table 1 for all types of irradiation (gamma rays, helium ions and carbon ions) in the case of pure DNA, DNA+PtTC and DNA+PtNP samples.

In order to make a quantitative analysis of the effects of the radiosensitizers, the sensitizing factors – SF – are defined as the ratio of the number of SSB (respectively DSB) per plasmid and per Gray induced in DNA loaded with platinum divided by the number of SSB (respectively DSB) per plasmid and per Gray obtained in pure DNA (see table 1 - column 4):

$$SF_{(SSB,DSB)} = \frac{m_{SSB,DSB,(Pt)}}{m_{SSB,DSB,(Pt=0)}}$$

Table 1. Yields of SSB and DSB (slopes *m*) induced by gamma rays (LET = 0.2 keV.μm^{-1}), He^{2+} ions (LET = 2.3 keV.μm^{-1}), C^{6+} ions from entrance (LET = 13 keV.μm^{-1}) and C^{6+} ions from SOBP (LET = 110 keV.μm^{-1}) in pure DNA, DNA + PtTC and DNA+ PtNP (with or without DMSO).

Irradiation source (LET)	Samples	Induction of breaks (per plasmid per gray)		Sensitizing Factor		$\frac{m_{SSB}}{m_{DSB}}$
		m_{SSB} $\times 10^{-4}$	m_{DSB} $\times 10^{-5}$	SF_{SSB}	SF_{DSB}	
Gamma ^{60}Co (LET = 0.2 keV/μm)	***Pure DNA***	*86 (±1.9)*	*5.2 (±0.2)*			*165*
	DNA+PtTC	139.5 (±2.4)	7.9 (±0.3)	1.6 (±0.1)	1.5 (±0.2)	177
	DNA+PtNP	120.1 (±1.7)	7.8 (±0.3)	1.4 (±0.1)	1.5 (±0.1)	154
	DNA+PtNP+DMSO	8.1 (±0.2)	0.7 (±0.1)			
Ions He^{2+} (LET = 2.3 keV/μm)	***Pure DNA***	*57.4 (±0.5)*	*8.6 (±0.4)*			*67*
	DNA+PtTC	97.8(±0.8)	13.2 (±0.5)	1.7 (±0.1)	1.5 (±0.2)	74
	DNA+PtNP	69.3 (±1.2)	13.3 (±0.7)	1.2 (±0.1)	1.5 (±0.2)	52
Ions C^{6+} (LET = 13 keV/μm)	***Pure DNA***	*53.1 (±0.6)*	*8.6 (±0.5)*			*62*
	DNA+PtTC	88 (±1)	13.8 (±0.6)	1.7 (±0.1)	1.6 (±0.2)	64
	DNA+PtNP	73 (±3)	18.3 (±0.8)	1.4 (±0.1)	2.1 (±0.3)	40
	DNA+PtNP+DMSO	7.2 (±1.9)	1.7 (±0.1)			
Ions C^{6+} (LET = 110 keV/μm)	***Pure DNA***	*34.5 (±0.2)*	*9.5 (±0.1)*			*36*
	DNA+PtTC	50.9 (±2.6)	14.9 (±0.9)	1.5 (±0.1)	1.6 (±0.1)	29
	DNA+PtNP	43.2 (±2.3)	12.3 (±0.4)	1.3 (±0.1)	1.3 (±0.1)	35
	DNA+PtNP+DMSO	4.4 (±0.3)	1.6 (±0.1)			

3.2. Role of direct and indirect effects

Some experiments were performed with DNA containing PtNP in the presence of a radical scavenger (DMSO) (see figure 1 for gamma rays radiation). DMSO is an efficient scavenger, which, at this concentration (1 mol.L^{-1}), totally inhibits the activity of water radicals. It is used to quantify the role of $HO^{\cdot}$ radicals. It should be noted that the chlorine contained in the TE buffer is also a radical scavenger but, at this concentration, it only mimics the scavenging conditions of living cells [18]. The calculated yields of DNA breaks (in the presence of PtNP) due to direct and indirect effects with gamma rays and C^{6+} are reported in table 2. A large decrease in the strand breaks yields is observed. It confirms previous results of the group obtained with carbon ions [14, 17]. These measurements show that the enhancement of SSB and DSB is mostly mediated by $HO^{\cdot}$ radicals for more than 90%. The effect of direct processes, including the direct ionization of the biomolecule by electrons or ions produced in the medium, and the effect of low energy electrons attachment, is very low. It indicates that the early stage processes initiated by the radiation and amplified by sensitizers ends in the major part with the production of water radicals prior to the induction of biological damage. These early stage processes have been described elsewhere [14].

Table 2. Yields of SSB and DSB per plasmid and per gray due to direct and indirect effects induced by gamma irradiation (LET= 0.2 keV.µm^{-1}) and ions C^{6+} irradiation (LET = 13 keV.µm^{-1}) in presence of PtNP.

		Gamma rays	C^{6+} ions
SSB	direct effect	7 %	10 %
	indirect effect	93 %	90 %
DSB	direct effect	9 %	9 %
	indirect effect	91 %	91 %

3.3. *Influence of LET on the sensitizing effect*

From table 1 we see that the induction of SSB decreases when LET increases. On the contrary, DSB increase with LET. As a result, the ratio m_{SSB} / m_{DSB} (table 1, last column) decreases when LET increases. This result confirms that ***the complexity of the damage increases with the ionizing density (LET) of the radiation*** [19]. The gamma rays have high energy and low LET resulting in dispersed ionizations events. Hence it is likely then that it will cause only single strand breaks in the DNA helix. In contrast, high LET charged particles such as carbon ions cause highly localized ionizations, which favor the induction of double-strand breaks.

This finding is also observed in the presence of sensitizers (PtTC and PtNP). In order to better compare the efficiencies of PtTC and PtNP, we report in figure 2 the sensitizing factors obtained with different types of radiation. For all types, the sensitizing factor of PtTC is close to 1.5 (± 0.1) for SSB and DSB (SF_{SSB} and SF_{DSB}). For PtNP, the sensitizing factors vary from 1.3 up to 2.1.

The goal of radiotherapy and radiosensitization is to eradicate cancer cells. We thus focus the discussion on the induction of complex molecular damage (DSB) which are the most lethal for the cells. As shown in figure 2, the efficiency of platinum molecules and nanoparticles (PtTC and PtNP) to induce complex damage –DSB– is found to be close to 1.5 on average for incident photons and ions. With the number of DSB increasing with LET, it is clearly *the combination of ion radiation and radiosensitizers wich causes the highest rate of complex damage*.

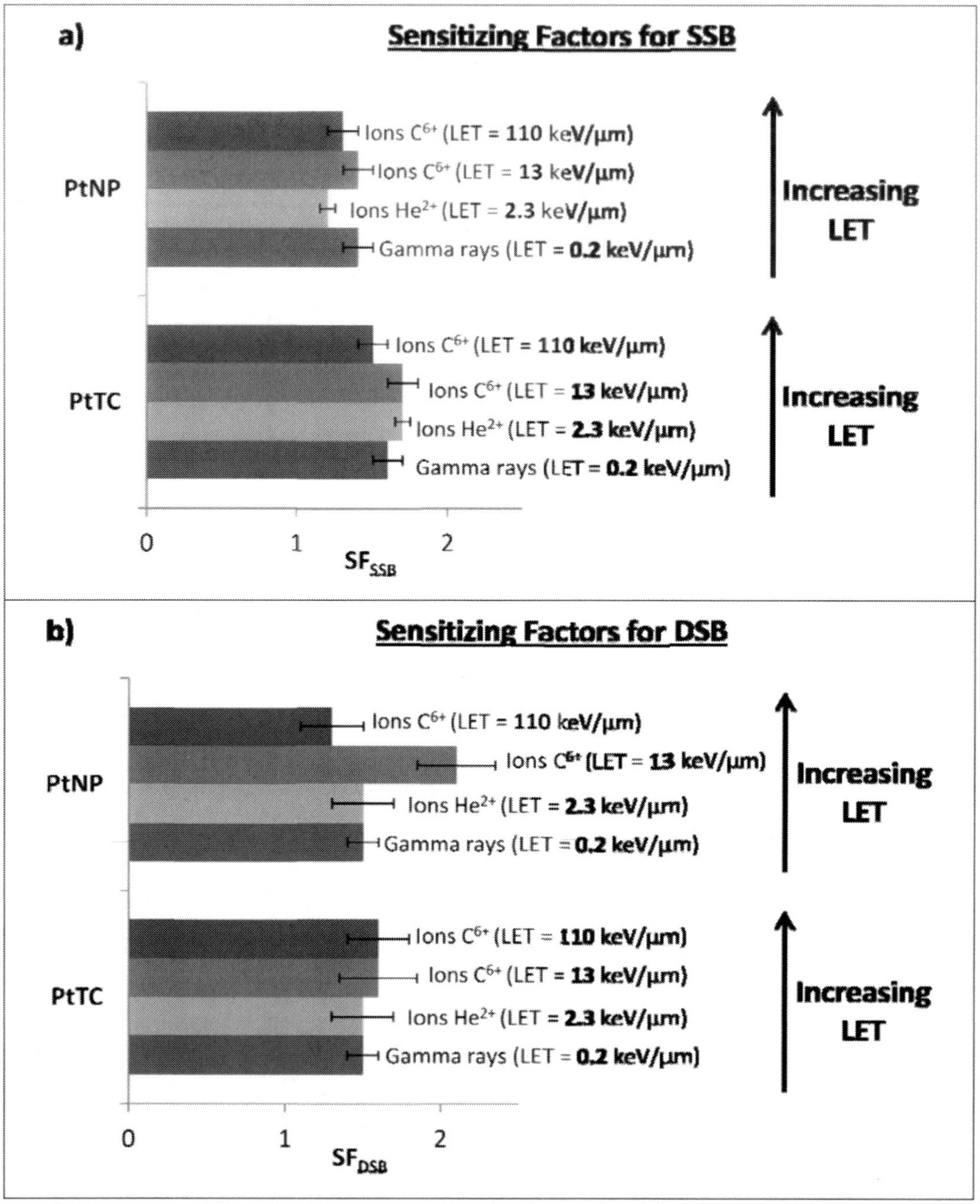

Figure 2. Sensitizing factors for SSB (a) and DSB (b) induced in DNA loaded with PtTC or PtNP for different LET (see values in table 1).

We notice that the best sensitizing effect is reached ($SF_{DSB} = 2.1$) when PtNP are combined with C^{6+} –LET = 13 keV.µm^{-1}– radiation. This implies that the nanoparticles are better electron emitters than isolated metal atoms (PtTC), because of auto-amplified electron cascades specific to nanoparticles [14].

In the case of incident helium ions (and gamma rays), there is no difference between PtNP and PtTC effects. This indicates that the nanoparticles are better activated by electrons emitted along the high LET particle track.

Interestingly, the efficiency of PtNP is getting lower (SF_{DSB} = 1.3) when the LET of carbon ions increases to 110 keV.µm^{-1}. This *decrease of sensitization for high LET carbon ions* compared to medium LET carbon ions can be *attributed to the well known mechanisms of HO˙ recombination.* Indeed, the production of hydroxyl radicals in the nanometric volume around the nanoparticles is expected to increase because of the ionizing density of the incident particles and because of the amplification induced by PtNP. The proximity of HO˙ species favors the recombination into less reactive H_2O_2. As a result, the induction of molecular damage decreases due to the decreasing HO˙ concentration. This phenomenon has been observed for cells irradiated by high LET radiation such as Argon ions [20-22]. In the presence of PtNP, this saturation occurs for lower LET [17].

4. Conclusion

The present experiments show that the presence of heavy atoms, in the form of salts or nanoparticles, induces a strong amplification of molecular damages (SSB, DSB) when gamma rays or fast ions (at different LET values) are used as ionizing radiation. The induction of complex damage being higher for high LET radiations, the most advantageous solution to enhance cell death is certainly the combination of sensitizers with fast ions. Finally, even though, the amplification at the Bragg peak was found lower than at the entrance, this factor is close to 1.3 which corresponds to an amplification of 30% of the molecular damage.

The use of a radical scavenger in some experiments confirms that the induction and the amplification of the molecular damage induced by incident ions and photons are mostly related to the production of free radicals in the medium (indirect processes). The direct processes (including direct breaks induced by electrons or photons) contribute with less than 10%.

The sensitization induced by nanoparticles is close to the effects of the salts (except for specific conditions of irradiation with carbon ions of medium LET). But, the main advantage of the NP stems from the possibility to functionalize their surface with tumor targeting molecules [3, 4]. The use of nanoparticles can thus improve the selective enrichment of cancer cells in sensitizing agents.

Finally, the combination of radiosensitizers and atomic ions could significantly improve the therapeutic index of radiotherapy. Regarding the results presented here, there is a strong need to develop and test new nanoparticles, different in composition, size and shape. There is also a strong need for molecular simulations to interpret and predict their effects with a view to better controlling the association of nanoparticles with fast ion radiation. In the end, the choice from a large number of radiosensitizers would open up prospects for personalized cancer treatment.

References

[1] Kobayashi K, Usami N, Porcel E, Lacombe S and Le Sech C 2010 Enhancement of radiation effect by heavy elements *Mutation Research-Reviews in Mutation Research* **704** 123-31

[2] Laster B H, Thomlinson W C and Fairchild R G 1993 Photon-Activation of Iododeoxyuridine - Biological Efficacy of Auger Electrons *Radiat. Res.* **133** 219-24

[3] Kong T, Zeng J, Wang X P, Yang X Y, Yang J, McQuarrie S, McEwan A, Roa W, Chen J and Xing J Z 2008 Enhancement of radiation cytotoxicity in breast-cancer cells by localized attachment of gold nanoparticles *Small* 4 1537-43

[4] Li X, Zhou H, Yang L, Du G, Pai-Panandiker A S, Huang X and Yan B 2011 Enhancement of cell recognition in vitro by dual-ligand cancer targeting gold nanoparticles *Biomaterials* **32** 2540-5

[5] Hainfeld J F, Slatkin D N and Smilowitz H M 2004 The use of gold nanoparticles to enhance radiotherapy in mice *Physics in Medicine and Biology* **49** N309-N15

[6] Mesbahi A 2010 A review on gold nanoparticles radiosensitization effect in radiation therapy of cancer *Reports of Practical Oncology & Radiotherapy* **15** 176-80

[7] Mowat P., Mignot A., Rima W., Lux F., Tillement O., Roulin C., Dutreix M., Bechet D., Huger S., Humbert L., Barberi-Heyob M., Aloy M.T., Armandy E., Rodriguez-Lafrasse C., Le Duc G., Roux S. and Perriat P. 2011 In vitro radiosensitizing effects of ultrasmall gadolinium based particles on tumour cells *Journal of Nanoscience and Nanotechnology* **11** 7833

[8] Butterworth K T, Wyer J A, Brennan-Fournet M, Latimer C J, Shah M B, Currell F J and Hirst D G 2008 Variation of strand break yield for plasmid DNA irradiated with high-Z metal nanoparticles *Radiat. Res.* **170** 381-7

[9] Kraft G 2003 ed. F Currell: Kluwer Academic Publishers Chapter 5: Radiobiological effects of highly charged ions

[10] Bridot J-L, Dayde D, Riviere C, Mandon C, Billotey C, Lerondel S, Sabattier R, Cartron G, Le Pape A, Blondiaux G, Janier M, Perriat P, Roux S and Tillement O 2009 Hybrid gadolinium oxide nanoparticles combining imaging and therapy *Journal of Materials Chemistry* **19** 2328-35

[11] Usami N, Furusawa Y, Kobayashi K, Lacombe S, Reynaud-Angelin A, Sage E, Wu T D, Croisy A, Guerquin-Kern J L and Le Sech C 2008 Mammalian cells loaded with platinum-containing molecules are sensitized to fast atomic ions *International Journal of Radiation Biology* **84** 603-11

[12] Jennette K W, Lippard S J, Vassiliades G A and Bauer W R 1974 Metallointercalation reagents. 2-hydroxyethanethiolato(2,2',2'-terpyridine)-platinum(II) monocation binds strongly to DNA by intercalation *Proc Natl Acad Sci U S A* **71** 3839-43

[13] Belloni J, Mostafavi M, Remita H, Marignier J L and Delcourt M O 1998 Radiation-induced synthesis of mono- and multi-metallic clusters and nanocolloids *New Journal of Chemistry* **22** 1239-55

[14] Porcel E, Liehn S, Remita H, Usami N, Kobayashi K, Furusawa Y, Le Sech C and Lacombe S 2010 Platinum nanoparticles: a promising material for future cancer therapy? *Nanotechnology* **21** 7

[15] Le Sech C, Takakura K, Saint-Marc C, Frohlich H, Charlier M, Usami N and Kobayashi K 2001 Enhanced strand break induction of DNA by resonant metal-innershell photoabsorption *Canadian Journal of Physiology and Pharmacology* **79** 196-200

[16] Spotheim-Maurizot M, Charlier M and Sabattier R 1990 DNA Radiolysis by Fast-Neutrons *International Journal of Radiation Biology* **57** 301-13

[17] Usami N, Kobayashi K, Furusawa Y, Frohlich H, Lacombe S and Le Sech C 2007 Irradiation of DNA loaded with platinum containing molecules by fast atomic ions C^{6+} and Fe^{26+} *International Journal of Radiation Biology* **83** 569-76

[18] Jones G D D, Milligan J R, Ward J F, Calabrojones P M and Aguilera J A 1993 Yield of Strand Breaks as a Function of Scavenger Concentration and Let for Sv40 Irradiated with He-4 Ions *Radiat. Res.* **136** 190-6

[19] Newman H C, Prise K M, Folkard M and Michael B D 1997 DNA double-strand break distributions in X-ray and alpha-particle irradiated V79 cells: Evidence for non-random breakage *International Journal of Radiation Biology* **71** 347-63

[20] Scholz M and Kraft G 1996 Track structure and the calculation of biological effects of heavy charged particles *Advances in space research : the official journal of the Committee on Space Research (COSPAR)* **18** 5-14

[21] Taucher-Scholz G, Heilmann J and Kraft G 1996 Induction and rejoining of DNA double-strand breaks in CHO cells after heavy ion irradiation *Advances in space research : the official journal of the Committee on Space Research (COSPAR)* **18** 83-92

[22] Gervais B, Beuve M, Olivera G H and Galassi M E 2006 Numerical simulation of multiple ionization and high LET effects in liquid water radiolysis *Radiation Physics and Chemistry* **75** 493-513

Journal of Physics: Conference Series **373** (2012) 012007 doi:10.1088/1742-6596/373/1/012007

Aggregation effects in proton collisions with water dimers

A. Ravazzani, L. F. Errea, L. Méndez and I. Rabadán

Laboratorio Asociado al CIEMAT de Física Atómica y Molecular en Plasmas de Fusión. Departamento de Química, Universidad Autónoma de Madrid, Madrid-28049, Spain.

E-mail: ismanuel.rabadan@uam.es

Abstract. Charge transfer cross sections in proton collisions with water dimers are calculated using an *ab initio* method based on molecular orbitals of the system. Results are compared with their counterpart in proton-water collisions to gauge the importance of intermolecular interactions in the cross sections.

1. Introduction

Ion collisions with water molecules are basic processes in hadron therapy since they give rise to ionization and fragmentation processes that produce electrons, ions and radicals, which lead to cellular damage by interaction with DNA. Detailed knowledge of ionizing reactions in ion-water collisions is also required for the simulation of the ion motion in the cell medium, where the simulation of the cell conditions requires to consider the interaction of ion projectiles with liquid water. However, previous theoretical and experimental work has focused on ion collisions with gas-phase water.

In order to gauge the importance of the interaction between water molecules on the relevant cross sections, we have considered proton collisions with the water dimer. To our knowledge, no other work has studied ion collisions with this species, and only Bouchiha *et al* [1] carried out calculations for electron-water dimer collisions. A different approach has been applied by Champion [2], who has employed the so-called Polarizable Continuum Model [3] to simulate the interaction of the water molecule involved in electron collisions with the liquid environment. In the present work, we have evaluated total cross sections for charge transfer (CT),

$$H^+ + (H_2O)_2 \rightarrow H + (H_2O)_2^+ \,, \tag{1}$$

and ionization (ION),

$$H^+ + (H_2O)_2 \rightarrow H^+ + (H_2O)_2^+ + e^- \,, \tag{2}$$

reactions in the collision energy range $1 \leq E \leq 40$ keV. We have applied the semiclassical method employed in reference [4] for ion collisions with water molecules. This method is based on the use of the independent electron approximation and the expansion of the one-electron wave function in a basis set of molecular orbitals. In our treatment we explicitly take into account the anisotropy of the molecular target by describing the interaction of the active electron with the molecular core by means of a multicentric pseudopotential. Atomic units are employed unless otherwise indicated.

2. Theory

At the collision energies considered in this work, the impact parameter and the Franck-Condon approximations can be applied. In the impact parameter approximation [5] the projectile follows rectilinear trajectories $\boldsymbol{R} = \boldsymbol{b} + \boldsymbol{v}t$, with impact parameter $\boldsymbol{b}$ and velocity $\boldsymbol{v}$; the motion of the active electron is described quantum-mechanically by the wave-function $\psi(\boldsymbol{r}, t)$ solution of the semiclassical equation [5]:

$$\left(h - \mathrm{i}\frac{\partial}{\partial t}\right)\psi(\boldsymbol{r}, t; \boldsymbol{b}, \boldsymbol{v}, \boldsymbol{\rho}) = 0\,, \tag{3}$$

where the components of vector $\boldsymbol{\rho}$ are the target nuclear coordinates, and $\boldsymbol{r}$ are the electronic coordinates, with the coordinates origin on the target center of mass. In the Franck-Condon approximation, the nuclei of the molecular target are assumed to remain fixed at their equilibrium positions, $\boldsymbol{\rho}_0$, while the non-adiabatic transitions leading to reactions (1) and (2) take place. The equilibrium geometry of the water dimer has been taken from [6].

Equation (3) is solved along each projectile trajectory as described in [4]. In particular, we explicitly consider the motion of the active electron in the field created by the nuclei and the remaining electrons, with the fixed-nuclei Hamiltonian h given by:

$$h = -\frac{1}{2}\nabla^2 + V_{\mathrm{T}} + V_{\mathrm{P}}, \tag{4}$$

with V_{T} the electron-target interaction potential and V_{P} the electron-projectile one. V_{T} is a multi-center pseudo-potential representing the interaction of the active electron with the molecular ion $(H_2O)_2^+$, and V_{P} is the Coulomb potential. The expression of V_{T} is:

$$\begin{aligned} V_{(\mathrm{H_2O})_2^+} &= \sum_k -\frac{N_k - N_{ck}}{r_k} - \frac{N_{ck}}{r_k}(1 + \alpha_k r_k)\mathrm{e}^{-2\alpha_k r_k}|s><s| \\ &\quad -\gamma_k \frac{N_{ck}}{r_k} r_k\,(1+\beta)\,\mathrm{e}^{-2\beta r_k}[1 - |s><s|]\,, \end{aligned} \tag{5}$$

with k labeling the nuclei $\{O_1, O_2, H_1, H_2, H_3, H_4\}$, N_k is the atomic number of the atom k, and r_k the distance between the active electron and the nucleus k. γ_k is equal to 1 for the Oxygen atoms and 0 for the Hydrogen atoms. The expression (5) contains the parameters N_{ck}, α_k and β, which have been fitted in order to minimize the differences between the orbital energies in the pseudo-potential and the energies of the corresponding orbitals in the calculation with correlation-consistent polarized valence triple-zeta basis set (HF/aug-cc-pVTZ) of NIST [7]; these differences are smaller than 4×10^{-3} Hartree for each valence orbital.

The monoelectronic wavefunction ψ of equation (3) is expanded as:

$$\psi(\boldsymbol{r}, t; b, v, \Omega) = \sum_j a_j(t; b, v, \Omega)\phi_j \exp\left(-\mathrm{i}\int_0^t \mathrm{d}t'\,\lambda_j\right), \tag{6}$$

where Ω is the solid angle that defines the orientation of the target dimer with respect to the projectile trajectory, and $\lambda_j = (\boldsymbol{s}^{-1}\boldsymbol{h})_{jj}$, where $\boldsymbol{s}$ and $\boldsymbol{h}$ are the overlap and Hamiltonian matrices in the basis $\{\phi_j\}$, which is a set of "asymptotically frozen" molecular orbitals (MO), eigenvectors of the matrix of h in a basis of Gaussian type orbitals (GTOs) at $R \to \infty$. In this work, we have used the GTOs given in the complementary material of [4] for the oxygen and projectile atoms and those of [8] for the molecular hydrogens. In the dynamical calculation we have employed a basis of 201 MOs, 45 of which are asymptotically bound orbitals and the rest, with positive energy, yield a discretization of the ionization continuum. Given the large basis employed, we have not included translation factors in the expansion.

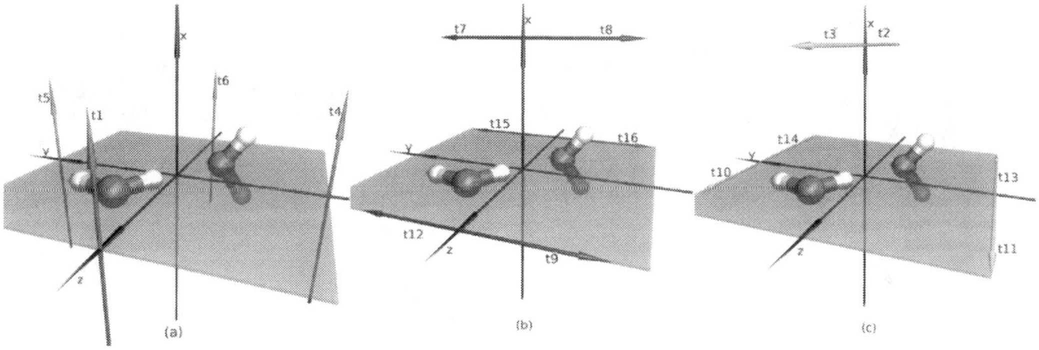

Figure 1. Equilibrium geometry of the water dimer (coordinates taken from [6]) and projectile-trajectory types used to obtain the orientation averaged cross sections of (12).

Substitution of (6) into (3) leads to a system of first-order differential equations whose solutions are the coefficients $a_j(t)$. In particular, for a collision where the electron is initially in the MO ϕ_i of $(H_2O)_2$, the probability p_{ij} for transition to other MO ϕ_j is obtained by solving this system of differential equations with the initial condition $a_j(t = -\infty) = \delta_{ij}$. This process is carried out for each occupied MO, while keeping closed the transitions to the other doubly occupied MOs. In this way, we obtain the monoelectronic probabilities:

$$p_{ij}(b, v, \Omega) = \lim_{t\to\infty} |\langle\phi_j(r)|\Psi\rangle|^2 = \lim_{t\to\infty} |a_j(t = \infty; b, v, \Omega)|^2 . \tag{7}$$

The monoelectronic probability for total CT from MO ϕ_i, p_i^{CT}, is the sum of those for transitions to the MOs located on the projectile and with negative energy. The transitions to MOs with positive energy lead to the corresponding ionization probability p_i^{ION}:

$$p_i^{\mathrm{CT}} = \sum_j p_{ij}^{\mathrm{CT}} \quad ; \quad p_i^{\mathrm{ION}} = \sum_j p_{ij}^{\mathrm{ION}} . \tag{8}$$

Many-electron probabilities are then obtained by applying the many-electron interpretation of reference [9], which is based on the Independent Event Model [10, 11]:

$$P^{\mathrm{CT}} = 2\sum_i p_i^{\mathrm{CT}}(1 - p_i^{\mathrm{ION}} - p_i^{\mathrm{SEC}}), \tag{9}$$

$$P^{\mathrm{ION}} = 2\sum_i p_i^{\mathrm{ION}}(1 - p_i^{\mathrm{ION}} - p_i^{\mathrm{SEC}}), \tag{10}$$

with i running over the target occupied MOs.

In order to compare our cross sections with experimental ones, we have obtained the orientation-averaged results:

$$\overline{\sigma}^{\mathrm{x}}(v) = \frac{1}{4\pi}\int \sigma^{\mathrm{x}}(v, \Omega)\mathrm{d}\Omega = \frac{1}{2}\int \mathrm{d}\Omega \int bP^{\mathrm{x}}(b, v, \Omega)\mathrm{d}b , \tag{11}$$

with x=CT or x=ION. The integration over the solid angle Ω is performed numerically using a 24-point Newton-Côtes formula detailed in [4] (and also used in [12]), which is a generalization of earlier work with more symmetric molecular targets [13]. In [4], it was shown that the CT cross sections depend considerably on the projectile-water molecule relative orientation and, following that work, we have carried out calculations with different orientations of the dimer with respect

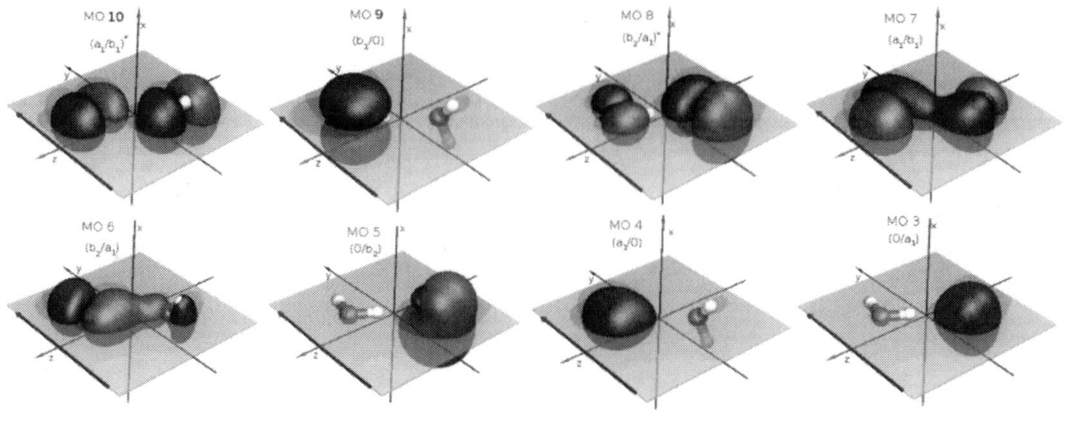

Figure 2. Molecular orbitals of the valence shell of $(H_2O)_2$. Indicated in the graphs are the symmetry in the C_{2v} point group of the MOs of the monomers with largest contributions to each MO of the water dimer. The blue arrow at the edge of the molecular plane is a t_{12} projectile trajectory.

to the projectile trajectory or, equivalently, for a set of proton trajectory orientations with respect to a fixed molecular target, as shown in figure 1. In this case, given the symmetry of the system, trajectories related by reflection on the YZ plane are equivalent, which reduces to 16 the number of different trajectory orientations required in the calculation from the original 24 (see [4]). With this, we obtain:

$$\overline{\sigma}^{\mathrm{x}}(v) \simeq \frac{1}{24}\left(\sum_{m=1}^{8} 2\sigma_m^{\mathrm{x}} + \sum_{m=9}^{16} \sigma_m^{\mathrm{x}}\right), \tag{12}$$

where m labels a particular trajectory orientation of figure 1. The right-most integral in (11) is performed in a grid of about 30 values of b, which means that, for any given projectile energy and each entrance channel, the calculation of $\overline{\sigma}^{\mathrm{x}}(v)$ required almost 500 projectile trajectories.

3. Results and discussion

Figure 2 shows the target MOs considered in this work. The arrow perpendicular to axis Z at the edge of the molecular plane corresponds to a projectile trajectory t_{12} that we will use to discuss the behaviour of the transition probabilities as functions of the impact parameter. The CT probabilities are shown in figure 3, at three collision energies. In all cases, we see that the transitions from MO 10 dominate the CT process with the maximum contribution from impact parameters in between 5 and 6 a_0. Also the CT probabilities from MOs 6, 4 and 7 are significant. The contribution of each MO to the CT process can be understood by looking at its shape and symmetry. In general, MOs with significant values near the projectile trajectory show larger CT probabilities because they have larger overlaps with projectile atomic orbitals, with the exception of MO 9. This exception is due to the fact that this particular MO is antisymmetric under reflection on the YZ plane, while the main CT channel, the Hydrogen 1s orbital, is symmetric. Accordingly, the overlap and interaction between entrance and exit orbitals vanish for trajectory t_{12}.

The orientation-dependent and orientation-averaged CT cross sections of equation (12) are presented in figure 4. We observe in this figure that the contribution of some orientations (t_{13}, for example) is up to three times larger than others (t_{14}), which underlines the importance of correctly accounting for the anisotropy of the target in the calculations. Incidentally, these

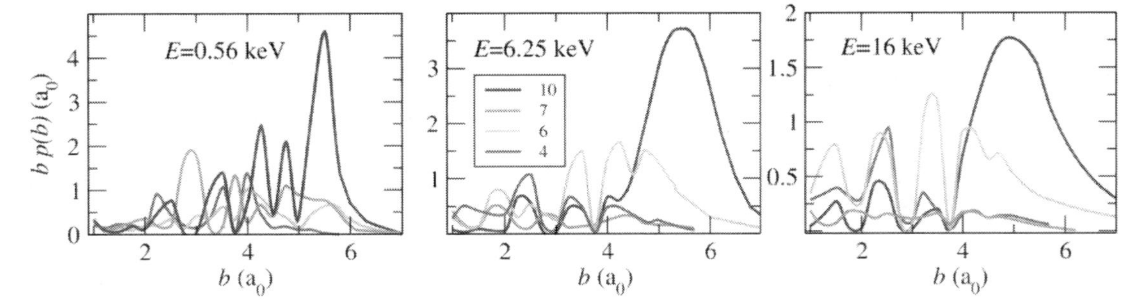

Figure 3. One-electron CT opacity functions, $bp_i^{\mathrm{CT}}(b)$ [equation (12)], as functions of the impact parameter for projectile trajectories t_{12} and for different projectile energies indicated in the panels. Line labels identify the initial MO ϕ_i (the entrance channel).

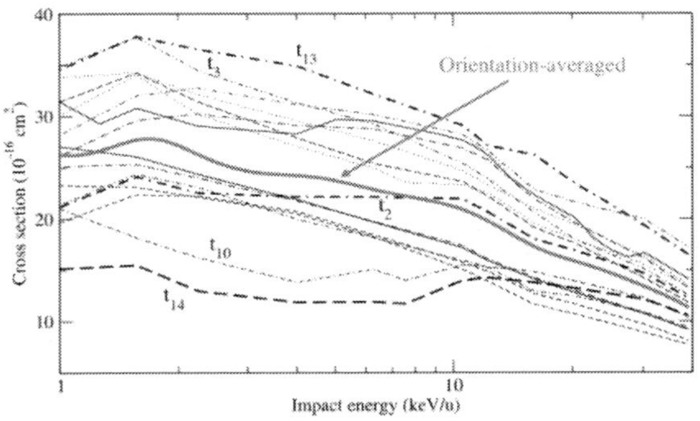

Figure 4. CT total cross sections for different trajectory orientations (see figure 1) and orientation-averaged CT total cross section evaluated using equation (12).

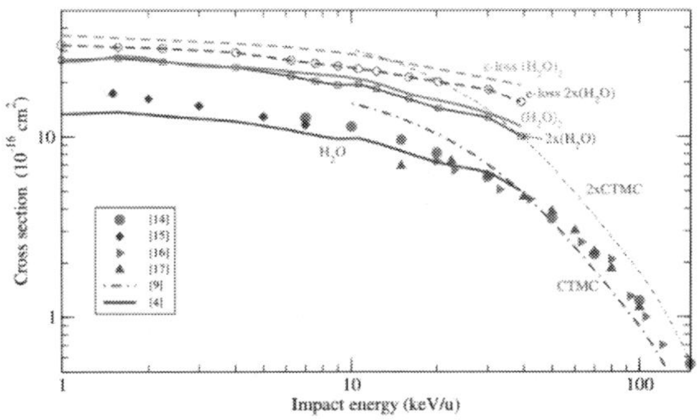

Figure 5. CT and electron loss cross sections in proton collisions with water dimers and with water monomers. Lines correspond to the present calculations and those of Gabás *et al* [4] and Illescas *et al* [9]. Solid symbols refer to CT experimental results of Rudd *et al* [14], Greenwood *et al* [15], Gobet *et al* [16], Luna *et al* [17].

results also help to identify a single trajectory type that leads to similar cross sections than the orientation-averaged one, t_2 in the present case, which could be useful in further simplified treatments.

The hydrogen-bond interaction in the water dimer produces both a shift in the monomer orbital energies and a redistribution of the electronic clouds. These two effects can influence the cross sections for the ionizing reactions (1) and (2), and they are included in this calculation through the target potential (5) that, as mentioned before, supports MOs whose energies are very close to those obtained in a standard SCF calculation. In this respect, we must point out that one of the features correctly reproduced in our calculations is that the CT channel lies below the highest occupied MO (HOMO) in H^+ collisions with the water dimer and not with the monomer.

In figure 5, we compare the present orientation-averaged cross sections with the corresponding ones in collisions of protons with two non-interacting water molecules, which, in the simplest approximation, are twice the ones obtained in collisions with a single water molecule. Figure 5 also includes de proton-water CT results of Gabás *et al* [4], evaluated using a method similar to the one employed here, classical (CTMC) results of Illescas *et al* [9] and experimental data [14, 15, 16, 17]. We plot the cross section for electron-loss, which is the sum of ionization and CT cross sections [reactions (1) and (2)], because, in general, the separation of CT and ionization fluxes is difficult when using molecular expansions. The figure shows a relatively small (about 20%) increase of the present electron-loss cross section with respect to that obtained from the proton-water result of reference [4]. Considering that the number of MOs with positive energy in the present work is about twice the one used in $H^{+}+H_2O$ collisions, we attribute that difference to the limited description of the continuum in [4]. On the other side, the MOs employed to describe the CT channels in collisions with both monomer and dimer targets are the same. There is remarkably good agreement between the present CT cross section and twice the result with the monomer, which indicates that the consequences on the electronic structure and geometry modifications due to the hydrogen-bond in the water dimer has no effect in the CT total cross section at the energies considered in this work.

Acknowledgments
This work has been supported by the project ENE2007-62934 of the Secretaría de Estado de Investigación, Desarrollo e Innovación (Spain). Allocation of computational time at the CCC of the Universidad Autónoma de Madrid is gratefully acknowledged.

References

[1] Bouchiha D, Caron L G, Gorfinkiel J D and Sanche L 2008 *J. Phys.* B **41** 045204
[2] Champion C 2010 *Phys. Med. and Biol.* **55** 11
[3] Tomasi J, Mennucci B and Cammi R 2005 *Chem. Rev.* **105** 2999–3094
[4] Gabás P M M, Errea L F, Méndez L and Rabadán I 2012 *Phys. Rev.* A **85** 012702
[5] Bransden B H and McDowell M H C 1992 *Charge Exchange and the Theory of Ion-Atom Collisions* (Oxford: Clarendon)
[6] Klopper W, van Duijneveldt-van de Rijdt J G C M and van Duijneveldt F B 2000 *Phys. Chem. Chem. Phys.* **2** 2227
[7] Russell D Johnson III (ed) 2011 *NIST Computational Chemistry Comparison and Benchmark Database* (NIST Standard Reference Database Number 101 Release 15b, August 2011,) URL `http://cccbdb.nist.gov/`
[8] Widmark P O, Malmqvist P and Roos B 1990 *Theor. Chim. Acta* **77** 291
[9] Illescas C, Errea L F, Méndez L, Pons B, Rabadán I and Riera A 2011 *Phys. Rev.* A **83** 052704
[10] Crothers D S F and McCarroll R 1987 *J. Phys.* B **20** 2835
[11] Janev R K, Solov'ev E A and Jakimovski D 1995 *J. Phys.* B **28** L615–L620
[12] Cabrera-Trujillo R, Deumens E, Ohrn Y, Quinet O, Sabin J R and Stolterfoht N 2007 *Phys. Rev.* A **75** 052702
[13] Errea L F, Gorfinkiel J D, Macías A, Méndez L and Riera A 1997 *J. Phys.* B **30** 3855–3872
[14] Rudd M E, Goffe T V and Itoh A 1985 *Phys. Rev.* A **32** 2128
[15] Greenwood J B, Chutjian A and Smith S J 2000 *Astrophys. J.* **529** 605–609
[16] Gobet F, Eden S, Coupier B, Tabet J, Farizon B, Farizon M, Gaillard M J, Carré M, Ouaskit S, Märk T D and Scheier P 2004 *Phys. Rev.* A **70** 062716
[17] Luna H, de Barros A L F, Wyer J A, Scully S W J, Lecointre J, Garcia P M Y, Sigaud G M, Santos A C F, Senthil V, Shah M B, Latimer C J and Montenegro E C 2007 *Phys. Rev.* A **75** 042711

Damage of DNA by Low Energy Electrons (< 3 eV)

Ilko Bald[1,2], Eugen Illenberger[1], and Janina Kopyra[3]

[1]Institut für Chemie-Physikalische und Theoretische Chemie, Freie Universität Berlin, Takustrasse 3, D-14195 Berlin
[2]present address: Department of Chemistry - Physical Chemistry, University of Potsdam, Karl-Liebknecht-Straße 24-25, 14476 Potsdam-Golm, Germany
[3]Chemistry Department, Siedlce University, 3 Maja 54, 08-110 Siedlce, Poland

E-mail: bald@uni-potsdam.de

Abstract. Recent experiments on low energy electron attachment to DNA and its components in the condensed phase and in the gas phase are reviewed and analysed. From different condensed phase experiments the sensitivity of DNA towards low energy electrons is well documented and strand breaks in DNA are observed at subexcitation energies (< 3 eV) and also in ultrafast electron transfer experiments involving electrons in presolvated states. Gas phase experiments indicate that all building blocks of DNA (the nucleobases, the sugar and the phosphate moiety) undergo resonant dissociative electron attachment (DEA) in the subexcitation regime which may ultimately lead to strand breaks. From very recent gas phase experiments on an entire nucleotide it can be concluded that most strand breaks result from direct electron attachment to the DNA backbone, but also initial electron capture by the nucleobase following electron transfer to the backbone contributes.

1. Introduction

It has been widely accepted by now that low energy electrons (LEEs) play a crucial role in radiation damage to biological material [1-3]. This is based on the fact that in the course of the interaction of high energy radiation (particles and/or photons) with condensed matter (e.g. a living cell) in the first step electrons are removed from any atomic or molecular quantum level of the material accessible at these energies. These *secondary electrons* can initiate further ionisation or excitation processes thereby being slowed down. Within femtoseconds following the initial interaction of the material with the high energy quantum, the energy distribution of these ballistic electrons extends to a few tens of eV [4]. Further inelastic scattering events slow the electrons down until they are thermalized before they reach some state of solvation, chiefly becoming by then chemically inactive species.

Before considering in more detail the action of high energy radiation to a living cell we shall briefly recall some elementary facts concerning the structure of DNA which represents the most vital component of the cell nuclei. The DNA is a biopolymer (figure 1) consisting of two chains (strands) which contain the four heterocyclic bases thymine (T), adenine (A), cytosine (C) and guanine (G). Each of these four bases is bound to the DNA backbone which in turn consists of sugar and phosphate units. Both strands are coupled through hydrogen bonds between base pairs in opposite positions in the two strands (figure 1). The architecture is such that adenine pairs with thymine (A-T) and guanine with cytosine (G-C) are resulting in the well known right handed double helix form. The DNA itself is surrounded by proteins and embedded in water, which is by far the most abundant component in a cell.

High energy quanta interacting with a living cell can obviously interact directly with DNA or with the surrounding water or proteins, either separately or in combination. Water as the most abundant component generates the reactive OH radical which in turn can attack DNA and its surrounding biomolecules (e.g. proteins). Two types of damage may be induced. *Direct damage* refers to energy deposition and subsequent reactions *directly* in the DNA and *indirect damage* refers to energy deposition into the environment of DNA and hence chiefly energy deposition to water thereby creating the reactive OH radical. It has been assumed that by far most of the indirect damage results from the attack of the highly reactive hydroxyl radical to DNA. It has been assumed that the damage of the genome by high energy radiation is about one third direct and two third indirect [5]. However, this notion has recently been questioned in the course of ultrafast electron transfer experiments to DNA [6] proposing that two thirds of the damage are direct and one third is indirect (see also below).

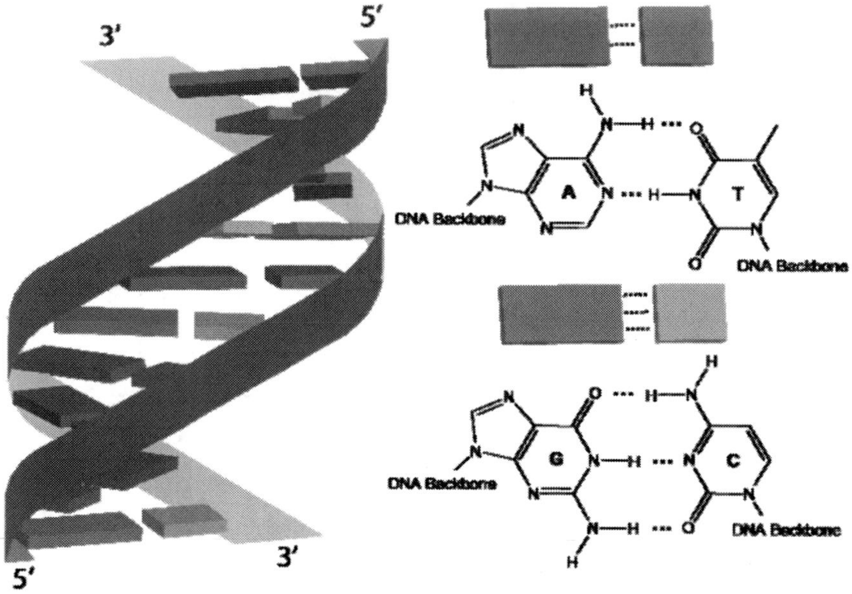

Figure 1. Helical structure of DNA and molecular structures of the base pairs adenine (A) – thymine (T) and guanine (G)- cytosine (C) connected by hydrogen bonding (from reference [3]).

The particular role of ballistic LEEs became directly obvious by a series of seminal experiments performed by Sanche and coworkers [7] who irradiated plasmid DNA by a well-defined electron beam at variable energy in the range 3-20 eV. They observed the appearance of single strand breaks (SSBs) and double strand breaks (DSBs) also in the range below the ionisation threshold with the efficiency showing a resonant behaviour with respect to the electron energy. Moreover, in the energy range below the level of electronic excitation (< 3 eV) strand breaks were also detected within structured resonances in this case, however, only SSBs [8]. From the resonant structure of the damage profile it was proposed that resonant electron capture could occur at particular molecular components and sites of the DNA and thus may be the initial step towards the observed strand breaks. It is well known that at low energies, preferentially below the level of electronic excitation, electrons can be captured by molecules thereby creating transient negative ions (TNIs) which could subsequently decompose into several fragments (dissociative electron attachment, DEA) [9]. This means that a chemical bond can be cleaved by electrons at energies well below the corresponding bond dissociation energy and sometimes already at threshold (0 eV). Energetically, such a low energy bond-breaking process becomes accessible provided that the electron binding energy on the negatively charged fragment is sufficiently large to locally increase the available energy. By that mechanism, therefore, the secondary ballistic electrons, on their way to be slowed down (but before getting solvated) can be captured by specific molecular sites within DNA and thereby trigger the chemical reactions that can eventually lead to cell apoptosis.

The sequence of reactions occurring along the ionisation track in the molecular network of a living cell may be divided into three major stages: (1) the *physical* stage which refers to processes and reactions in the time window from 10^{-15}s (fs) to 10^{-12}s (ps) after the primary interaction. This time window includes electronic excitation and ionisation with subsequent bond ruptures along repulsive potential energy surfaces, the formation of OH radicals and formation of abundant numbers of secondary electrons, (2) the *chemical stage,* which includes molecular relaxation and reorganisation including multiple bond cleavages and the subsequent formation of new molecules. Such reactions typically occur in the time window between picoseconds (10^{-12}s) to microseconds (10^{-6} s) (metastable decays), and (3) the *biological stage,* which finally refers to the alterations occurring on a longer time scale, including the overall response of the system which could extend over a time scale of several years and more.

Among this sequence of processes and reactions the fast and ultrafast initial events (driven to a large extent by LEEs) are considered as important and decisive steps for the evolution of the biomolecular system and hence for the chemical and biological effects occurring on a much longer time scale. In the following sections, we shall first describe a few experiments performed on DNA and its gas components in the condensed phase, and then, describe electron driven reactions on gas phase building blocks. From the latter experiments, the intrinsic properties of the DNA components with respect to LEE interaction can be studied. Together with larger units and the results from the condensed phase, a fairly detailed description of the molecular processes, in particular how low energy electrons damage DNA, can be obtained.

2. Results and discussion

2.1. Electron induced damage to DNA in the condensed phase

The landmark experiments by Sanche and coworkers demonstrated that LEEs can induce strand breaks (DSBs and SSBs) below the level of ionisation and also at subexcitaion energies, albeit in the latter case only SSBs. While the resonant profile of the damage efficiency suggested resonant capture of electrons as the initial step, the detailed mechanism leading to strand breaks was not obvious. For the molecular mechanism inducing SSBs it was theoretically predicted [10,11] that LEEs are captured by the DNA bases and transferred to the backbone thereby rupturing the connection between the phosphate and the sugar unit representing a strand break (figure 1). Different experiments on model DNA in fact revealed the particular sensitivity of the biopolymer towards LEEs.

Particularly interesting results with respect to radiation damage have been obtained by low energy photoelectron transmission spectroscopy (LEPET) applied to self-organized layers of short single and double stranded oligonucleotides of known size (15 units) and sequence [12]. In LEPET electrons are detected, which are photo-ejected from the metallic substrate, transmitted through the film and are finally escaping into vacuum. It was found that the oligonucleotides readily capture electrons within the energy spectrum covered by the photo-injection technique which is in the subexcitation range, and that the transmitted current significantly depends on the presence of guanine (G), *i.e.,* the number of G bases controls the capturing efficiency. These findings support the picture that the DNA bases act as antennae for the capturing of subexcitation electrons and that SSBs are induced via electron transfer to the backbone.

A high throughput method to study electron-induced damage in model DNA was established by immobilising single stranded short oligonucleotides on a gold surface in a microarray format and exposing them to electrons in the subexcitation energy range (1-3 eV) [13]. It was found that damage to single stranded oligomers (25 units) is detected at quite a low electron dosage of about 300 electrons per oligonucleotide [14] and that damage correlates linearly with the number of guanines [15] which in turn correlates with the results from LEPET mentioned above.

Journal of Physics: Conference Series **373** (2012) 012008 doi:10.1088/1742-6596/373/1/012008

Recent experiments on plasmid DNA in water using femtosecond (fs) laser techniques revealed the particular role of presolvated electrons in radiation damage [6]. Water is ionised and/or excited by the absorption of two UV photons creating OH radicals and electrons. Within a time window of about 100 fs the electrons reach a kind of presolvated state. It was claimed that SSBs and DSBs induced by electron transfer from this presolvated stage (reductive damage) is twice as efficient as that of strand breaks induced by the OH radicals (oxidative damage [16]). The important role of presolvated electrons in radiation damage has also been pointed out before [17,18].

More direct information on the molecular mechanisms following electron attachment to DNA and its components can be obtained from gas phase experiments as described below.

2.2. *Electron induced damage to gas phase building blocks of DNA*

2.2.1. DNA bases. The first gas phase experiments on DNA bases were carried out in 1998 in the Berlin laboratory by means of a crossed electron-molecular beam experiment [19]. Negative ions created by the interaction of the electron beam with the effusive molecular beam (resulting from thermal evaporation) were extracted from the reaction volume and analysed by means of a quadrupole mass spectrometer.

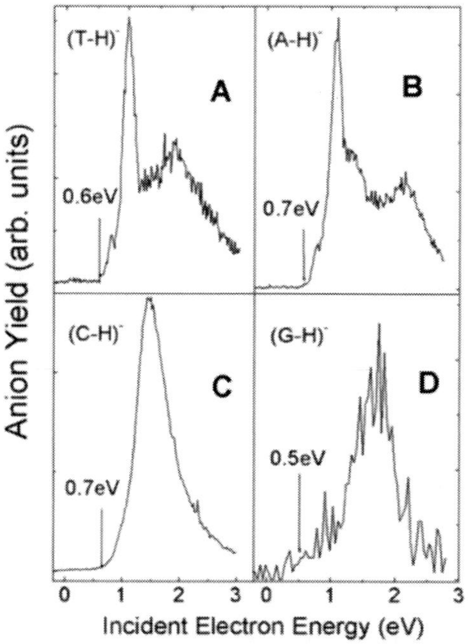

Figure 2. Low energy DEA in the 4 DNA bases leading to the loss of a neutral hydrogen atom. $(T\text{-}H)^-$ assigns the closed shell anion of thymine, formed by hydrogen loss from the transient anion (from reference [21]).

These earlier gas phase results clearly revealed the presence of low energy resonances in the DNA Bases T, C, G, A and U, namely in the energy range around 6-12 eV and in the energy range around 1 eV. While the negative ion fragments from T originating from the higher energy resonances indicated DEA reactions involving the opening of the cyclic structure, the low energy resonance yielded an ion at a mass close to that of the corresponding parent for each nucleobase (NB). In fact, it became soon obvious in experiments by the Innsbruck and Berlin laboratories [20,21] (see figure 2) that all NBs exhibit a low energy resonance associated with the loss of a neutral hydrogen atom according to the dissociative electron attachment (DEA) process

$$e^{-} + NB \rightarrow (NB\text{-}H)^{-} + H \qquad (1)$$

where $(NB\text{-}H)^{-}$ describes the closed shell anion formed by ejection of a neutral hydrogen radical from the target molecule. The reaction is already operative at sub-excitation energies and is energetically driven by the appreciable electron affinity of the (NB-H) radicals which in the case of T lies in the range between 3 and 4 eV.

More detailed studies in T revealed an unexpected bond and site selectivity for the hydrogen loss process in that only the N-H bonds are involved [21], and most remarkably, that the sharp peak in the $(N\text{-}H)^{-}$ ion yield located at 1 eV (figure 2) exclusively results in hydrogen loss from the N1 position, and the broader resonance peaking at 1.8 eV preferentially leads to hydrogen loss fom the N3 position [22]. For T coupled in the DNA network the N1 position is involved in the coupling to the sugar unit and the N3 position in hydrogen bonding to the complementary base A (see figure 1).

The general situation in the isolated DNA bases is then the presence of low energy resonances at subexcitation energies exclusively associated with the loss of a neutral hydrogen atom, and more or less structured resonance features in the energy range between 6 and 12 eV, which are associated with complex DEA reactions leading to the deterioration of the cyclic structure [19,20,23] (not shown here). To which degree these resonances are involved in SSBs or DSBs depends on how these properties propagate when the bases are covalently coupled within the DNA network. While the molecular mechanisms leading to DSBs are presumably very complex and not obvious so far, a clearer picture on how subexcitation electrons induce single strand breaks (SSBs) is emerging from the gas phase studies. As mentioned above, quantum chemistry studies [10,11] predict capture of a subexcitation electron by the DNA bases and transfer to the backbone, ultimately leading to the rupture of the C-O bond between the phosphate and the sugar moiety. The question then arises on the sensitivity of the sugar and the phosphate moiety (representing the DNA backbone) towards low energy electrons and hence their possible contributions to direct SSBs.

2.2.2. Components of the DNA backbone. The DNA backbone is known to consist of alternating phosphate units and the sugar 2-deoxy-D-ribose, with the nucleobases connected via the glycosidic C-N bond (see figure 3, top). Obviously the cleavage of any of the P-O-C bonds (connecting the phosphate with the sugar moiety) but also cleavage of the C4-C5 bond within the sugar unit would represent a strand break. In addition, a decomposition of the sugar unit within DNA would result in both strand breaks and nucleobase release. The sensitivity of the components of the backbone towards low energy electrons is hence of particular interest as decomposition reactions on this site of the DNA could act as one of the most direct mechanisms for single strand breaks.

Electron attachment experiments to different six- and five-membered isolated sugar molecules showed a surprisingly high sensitivity of these compounds towards LEEs already at energies close to zero eV. Electron capture is associated with remarkably complex reactions leading to the loss of one to several H_2O and H_2CO molecules [24-27].

A more realistic view on the behaviour of the sugar and phosphate moiety in DNA may be achieved by using tetraacetylribose (TAR) and the phosphoric acid esters dibutylphosphate (DBP) and triethylphosphate (TRP) as the corresponding surrogate (figure 3 bottom).

TAR possesses a five membered ring structure and no free hydroxyl group, and the acetyl groups represent polar moieties with an electron affinity similar to the phosphate group and the nucleobases. The electron attachment and fragmentation reactions in TAR can specifically be assigned to resonances which are located at particular sites within the molecule [28]. A group of heavier fragment anions is exclusively generated through resonances peaking at 1.6 – 1.8 eV (not shown here), but also a resonance close to zero eV yielding a variety of fragments, some of the ion yields are shown in figure 4. These fragment ion yield curves mirror the decomposition of the ring and are also reminiscent of the yield curves observed for bare sugar molecules.

The first gas phase results on DEA processes involving phosphoric acid derivatives were obtained on dibutylphosphate (DBP) and triethylphosphate (TEP) [29]. Both compounds are subject to DEA

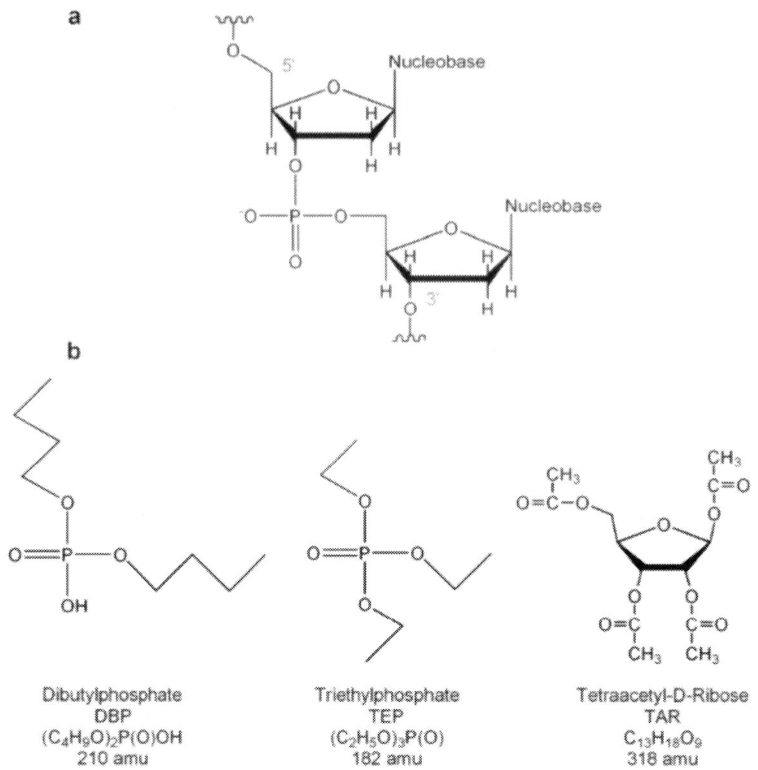

Figure 3. Molecular structure of the DNA backbone consisting of the sugar and the phosphate units (top), and molecular structures of dibutylphosphate (DBP), triethylphosphate (TEP) and tetraacetyl-D-ribose (TAR), which serve as surrogates for the phosphate or the sugar unit in DNA, respectively.

processes occurring at different energies, ranging from close to 0 eV up to 10 eV. The most important reaction in the context of DNA damage is the loss of a butyl group via C-O bond cleavage. This DEA reaction takes place within two broad resonances located at 2-4 eV and 7-10 eV (figure 5 left).

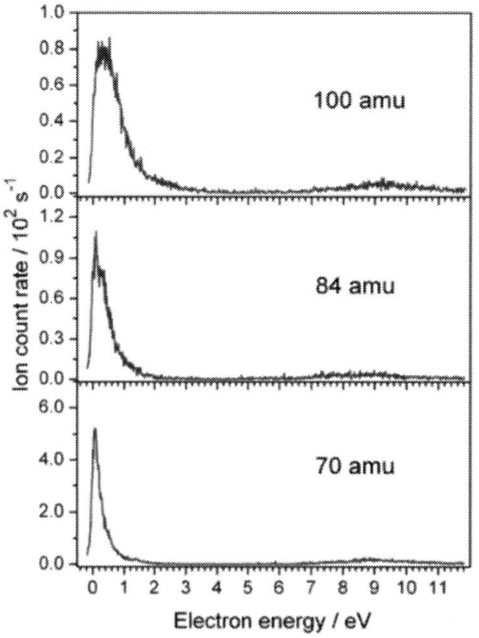

Figure 4. Fragment ions at 70 amu, 84 amu and 100 amu arising from DEA to tetraacetyl-D-ribose (TAR). These ions are produced close to 0 eV and within a weaker resonance located at 7 – 9 eV and are hence reminiscent to the ones obtained from DEA to bare D-ribose (adapted from reference [28]).

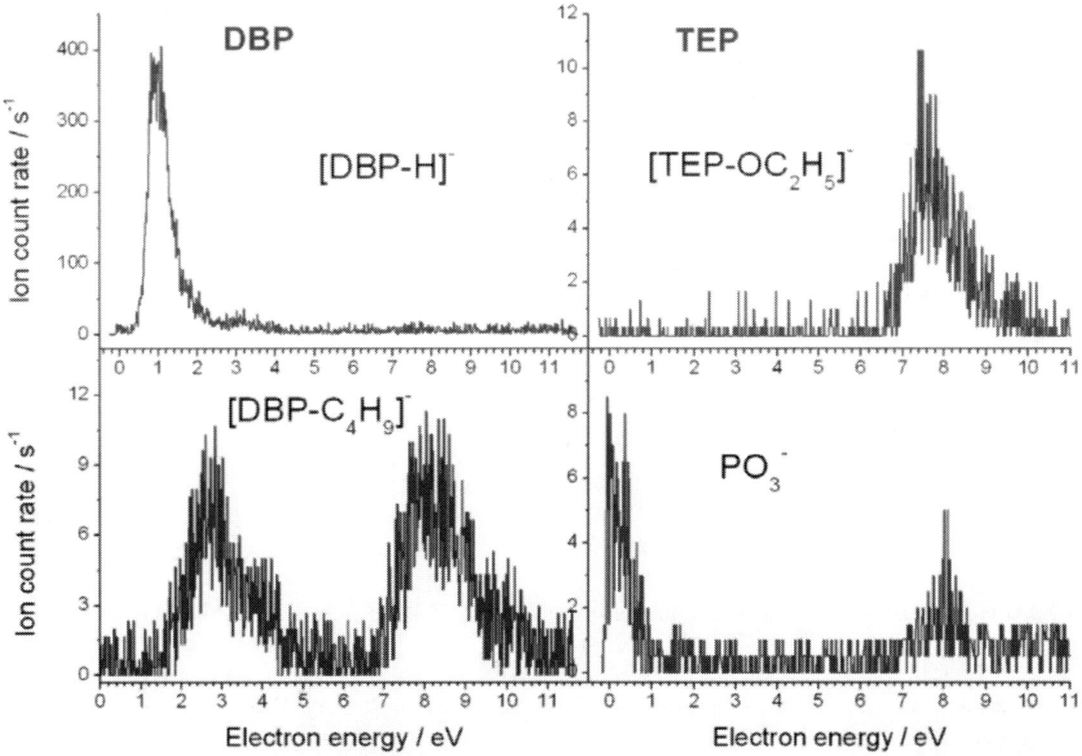

Figure 5. Ion yields of selected fragmentation products generated by DEA to dibutylphosphate (DBP, left) and triethylphosphate (TEP, right) (adapted from [29]).

A similar reaction was observed in TEP (figure 5 right), *i.e.*, the loss of an ethoxy radical due to P-O bond cleavage. It should be emphasised that for the phosphate group in DNA both reactions would correspond to a direct single strand break.

Another remarkable set of reactions takes place in both DBP and TEP close to zero eV and is associated with the excision of the negatively charged central phosphate unit (not shown here), in TEP only PO_3^- was observed at zero eV, driven by the unusually large electron affinity of PO_3 (4.95 eV). Obviously, the excision of these central units requires multiple bond cleavages which, in DNA, would certainly result in a strand break.

2.2.3. Larger DNA building blocks. Experiments on larger building blocks in the gas phase, namely thymidine (thymine coupled to sugar) [30] and D-ribose-5-phosphate [31] showed that the low energy resonances present in the separated molecules are still present in the covalently coupled compounds undergoing effective DEA reactions. From a very recent experiment on an entire nucleotide (2'-deoxycytidine 5′-monophosphate, dCMP) [32] it can be deduced that SSBs driven by subexcitation electrons predominantly occur by *direct* electron localisation at the backbone, but also electron transfer reactions involving initial electron localisation at the nucleobase are contributing. This can be concluded from the nature of the final ionic products and their resonance profiles which in turn carry the signature of the initial attachment process.

Acknowledgments
Work supported by the Polish Ministry of Science and Higher Education and the Deutsche Forschungsgemeinschaft (DFG).

References
[1] Sanche L 2005 *Eur. Phys. J.* D **35** 367
[2] Sanche L 2010 *Low-Energy Electron Interaction with DNA: Bond Dissociation and Formation of Transient Anions, Radicals and Radical Anions*, in *Radical and Radical Ion Reactivity in Nucleic Acid Chemistry* ed M M Greenberg (John Wiley & Sons)
[3] Baccarelli I, Bald I, Gianturco F A, Illenberger E and Kopyra J 2011 *Phys. Rep.* **508** 1
[4] Cobut V, Fongillo Y, Patau J P, Goulet T, Fraser M-J and Jay-Gerin J-P 1998 *Radiat. Phys. Chem.* **51** 229
[5] Michael B D and O'Neill P A 2000 *Science* **287** 1603
[6] Nguyen J, Ma Y, Luo T, Bristow R G, Faffray D A and Lu Q B 2011 *Proc. Natl. Acad. Sci. U.S.A. (PNAS)* **108** 11778
[7] Boudaïffa B, Cloutier P, Hunting D, Huels M A and Sanche L 2000 *Science* **287** 1658
[8] Martin F, Burrow P D, Cai Z, Cloutier P, Hunting D and Sanche L 2004 *Phys. Rev. Lett.* **93** 068101
[9] Illenberger E and Momigny J 1992 *Gaseous Molecular Ions. An Introduction to Elementary Processes Induced by Ionization* (Darmstadt: Steinkopff Verlag / New York: Springer-Verlag)
[10] Berdys J, Anusiewicz I, Skurski P and Simons J 2004 *J. Am. Chem. Soc.* **126** 6441
[11] Simons J 2006 *Acc. Chem. Res.* **39** 772
[12] Ray S G, Daube S S and Naaman R 2005 *Proc. Natl. Acad. Sci. U.S.A. (PNAS)* **102** 15
[13] Solomun T and Illenberger E 2004 *Chem. Phys. Lett.* **396** 448
[14] Solomun T and Sturm H 2007 *J. Phys. Chem.* B. **111** 10636
[15] Solomun T, Seitz H and Sturm H J. 2009 *Phys. Chem.* B **113** 11557
[16] von Sonntag C 2006 *Free Radical Induced DNA Damage* (Berlin: Springer-Verlag)
[17] Wang C R, Nguyen J, Lu Q-B 2009 *J. Am. Chem. Soc.* **131** 11320
[18] Sanche L 2009 *Nature* **416** 358
[19] Huels M A, Hahndorf I, Sanche L and Illenberger E 1998 *J. Chem. Phys.* **108** 1309
[20] Hanel G, Gstir B, Denifl S, Scheier P, Probst M, Farizon B, Farizon M, Illenberger E and Märk T D 2003 *Phys. Rev. Lett.* **90** 188104
[21] Abdoul-Carime H, Gohlke S and Illenberger E 2004 *Phys. Rev. Lett.* **92**, 168103
[22] Ptasinska S, Denifl S, Scheier P, Illenberger E and Märk T D 2005 *Angew. Chem. Int. Ed.* **117** 7101
[23] Denifl S, Ptasinska S, Probst M, Hrusak J, Scheier P and Märk T D 2004 *J. Phys. Chem.* A **108** 6562
[24] Ptasinska S, Denifl S, Scheier P, and Märk T D 2004 *J. Chem. Phys.* **120** 8505
[25] Bald I, Kopyra J and Illenberger E 2006 *Angew. Chem. Int. Ed.* **45** 4851
[26] Sulzer P, Ptasinska S, Zappa F, Mielewska B, Milosavljevic A R, Scheier P, Märk T D, Bald I, Gohlke S, Huels M A and Illenberger E 2006 *J. Chem. Phys.* **125** 044304
[27] Baccarelli I, Gianturco F A, Grandi A, Sanna N, Lucchese R R, Bald I, Kopyra J and Illenberger E 2007 *J. Am. Chem. Soc.* **129** 6269
[28] Bald I, Kopyra J, Dabkowska I, Antonsson E and Illenberger E 2007 *J. Chem. Phys.* **126** 074308
[29] König C, Kopyra J, Bald I and Illenberger E 2006 *Phys. Rev. Lett.* **97** 018105
[30] Ptasinska S, Denifl S, Gohlke S, Scheier P, Illenberger E and Märk T D 2006 *Angew. Chem. Int. Ed.* **45** 1893
[31] Bald I, Dabkowska I and Illenberger E 2008 *Angew. Chem. Int. Ed.* **47** 8518
[32] Kopyra J 2012 Phys. Chem.Chem. Phys. **14** 8287

Journal of Physics: Conference Series **373** (2012) 012009 doi:10.1088/1742-6596/373/1/012009

Electron interactions with positively and negatively multiply charged biomolecular clusters

Linda Feketeová

School of Chemistry, Bio21 Institute and ARC Centre of Excellence for Free Radical Chemistry and Biotechnology, The University of Melbourne, Victoria 3010, Australia

lfe@unimelb.edu.au

Abstract. Interactions of positively and negatively multiply charged biomolecular clusters with low-energy electrons, from ~ 0 up to 50 eV of electron energy, were investigated in a high resolution Fourier-Transform Ion Cyclotron Resonance mass spectrometer equipped with an electrospray ionisation source. Electron-induced dissociation reactions of these clusters depend on the energy of the electrons, the size and the charge state of the cluster. The positively charged clusters $[M_n+2H]^{2+}$ of zwitterionic betaines, $M=(CH_3)_2XCH_2CO_2$ ($X = NCH_3$ and S), do capture an electron in the low electron energy region (< 10 eV). At higher electron energies neutral evaporation from the cluster becomes competitive with Coulomb explosion. In addition, a series of singly charged fragments arise from bond cleavage reactions, including decarboxylation and CH_3 group transfer, due to the access of electronic excited states of the precursor ions. These fragmentation reactions depend on the type of betaine ($X = NCH_3$ or S). For the negative dianionic clusters of tryptophan $[Trp_9–2H]^{2-}$, the important channel at low electron energies is loss of a neutral. Coulomb explosion competes from 19.8 eV and dominates at high electron energies. A small amount of $[Trp_2–H–NH_3]^-$ is observed at 21.8 eV.

1. Introduction

Absorption of high-energy particles or radiation by a living cell is accompanied by the production of secondary reactive species that play a dominant role in the induction of radiation damage [1]. One of the most abundant secondary species are electrons, with the yields of around $5x10^4$ per MeV of the incident radiation, and kinetic energies < 20 eV [2]. Nano-scale insights into the ion beam cancer therapy also showed that the ionisation of the medium is the major source of energy loss of the ion projectile resulting in a shower of secondary electrons, with the large majority being produced below 30 eV, while the number of produced electrons with kinetic energies > 100 eV is negligible [3]. These electrons react with molecular components of the cell to form negative ions and radicals, which in turn may be crucial in the damage to the living organism. It has been shown that low-energy electrons can induce substantial damage, such as base and sugar modifications, base release, single strand breaks and double strand breaks [4]. Clearly the microscopic analysis of the whole scenario, and the fundamental understanding of electron interactions with biological molecules, is vitally important in order to provide more accurate predictions for ion beam cancer therapy, shielding of human space missions, or to predict the consequences of exposure to photon radiation.

Recently, there has been increased interest in the study of biomolecular clusters in order to understand the influence of the environment on the fragmentation of biomolecules. Neutral biomolecular clusters or biomolecules embedded in Helium droplets are investigated with low-energy electrons [5] or ion beams [6]. The advent of electrospray ionization (ESI) [7] has provided a new method for generating a wide range of cluster ions, including ionic clusters and biomolecular clusters. The study of biomolecular cluster ions has become an active research field, focused on their formation, structure and dissociation reactions [8,9]. Most studies on the dissociation reactions of biomolecular cluster ions have employed vibrational activation techniques such as collision-induced dissociation (CID) [8]. Only a few studies have focussed on the use of ion-electron reactions to promote the fragmentation [10] and even less is known about how dissociation of charged biomolecular clusters might depend on the energy of impinging electrons. The present manuscript is an overview of work recently published on electron interactions with positively and negatively multiply charged biomolecular clusters studied in a range of electron energies from ~ 0 up to 50 eV [11-13]. Various aspects of the formation, stability and fragmentation induced by low-energy electrons are discussed. Clusters close to the stability limit, where Coulomb repulsion of the charges within the cluster competes with attractive forces holding the cluster together, as well as large clusters are described. The novel aspect of the presented work is the investigation of ion-electron interactions as a function of the electron energy that has been explored for the first time with multiply charged biomolecular cluster ions.

2. Mass spectrometry

All experiments were carried out on a Finnigan- LTQ-FT (Thermo, Bremen, Germany) mass spectrometer equipped with an electrospray ionisation (ESI) source [14] described in detail elsewhere [15]. For details of the experiments overviewed in the current publication please refer to [11-13]. Briefly, the LTQ-FT mass spectrometer consists of: (i) a linear ion trap (LTQ); (ii) ion transfer optics; and (iii) a Fourier-transform ion cyclotron resonance (FT-ICR) mass analyser (figure 1). The desired ions produced via ESI can be trapped, isolated and subjected to CID in the LTQ at a He bath gas pressure of ca 5 x 10^{-3} Torr. For high-resolution mass analysis and electron-ion interactions, the ions are transferred via the ion optics transfer region (~2x10^{-7} Torr) into the FT-ICR cell at a pressure below 1.5x10^{-9} Torr. The FT-ICR cell is supplied with low-energy electrons produced by an indirectly heated emitter cathode located downstream of the FT-ICR cell. The energy of electrons is given by the potential difference between the emitter cathode and the grid positioned in front of the cathode, and can be varied from ~ 0 up to 100 eV.

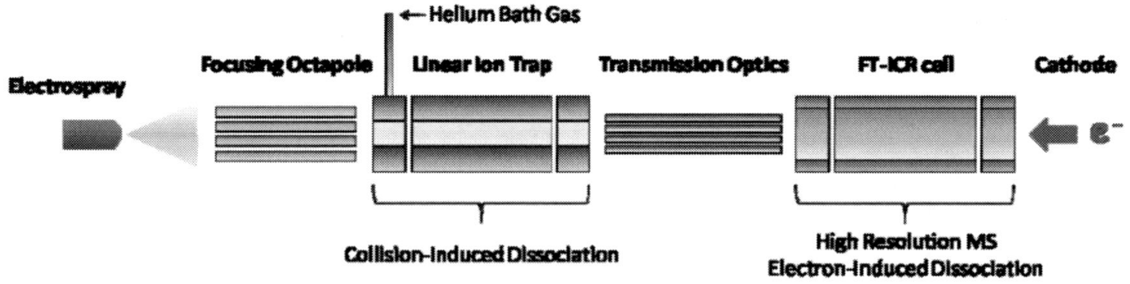

Figure 1. Schematic overview of the Finnigan- LTQ-FT mass spectrometer.

3. Results and discussion

3.1. Positively charged clusters of zwitterionic betaines

ESI of a solution of GB (figure 2a) and DMSA (figure 2b) yields a series of protonated and sodiated clusters of types $[M_n+H]^+$ and $[M_n+Na]^+$ as well as the multiply charged clusters $[M_n+2H]^{2+}$ and $[M_n+3H]^{3+}$. It was shown that these clusters are stabilized due to the fact that GB and DMSA are permanent zwitterions, as such clusters have not been observed for its isomer N,N-dimethylglycine methyl ester [16], which is lacking the zwitterionic structure. To promote the formation of large, higher charged clusters of DMSA, the concentration of DMSA in the ESI solution had to be increased 20 times (20 mM) in comparison to GB (1 mM). Additionally, the increase of the amount of acetic acid in the ESI solution from 1 % (case of GB) to 3 % helped towards higher intensity of DMSA multiply charged cluster ions, note the logarithmic scale of the intensity in the ESI mass spectrum of GB.

All the experimental conditions were varied in order to find out the size of the smallest multiply charged clusters formed in the ESI. The smallest doubly charged cluster was for both, GB and DMSA, the $[M_{13}+2H]^{2+}$. However, for the triply charged cluster of DMSA, the smallest cluster needed extra two neutrals to overcome Coulomb repulsion of the three charges, thus forming $[M_{26}+3H]^{3+}$ as the smallest triply charged cluster as oppose to $[M_{24}+3H]^{3+}$ of GB. This observation suggested a lower binding energy of DMSA neutrals within the cluster. Further investigation showed that the binding energy of DMSA neutrals in the cluster is 0.86 times smaller than that of GB [12]. The intriguing question that arose was "How are these multiply charged clusters held together in the gas phase?" PM3 semi-empirical and DFT calculations [16] suggested that the core of the doubly charged clusters might adopt the centrosymmetric dimer structure typically observed in the self-association of carboxylic acids [17]. Forming a doubly charged dimer core, the Coulombic repulsion of the charge of the two cations [18] can be overcome by attractive forces such as hydrogen bonding [19] and dipole-dipole interactions [20].

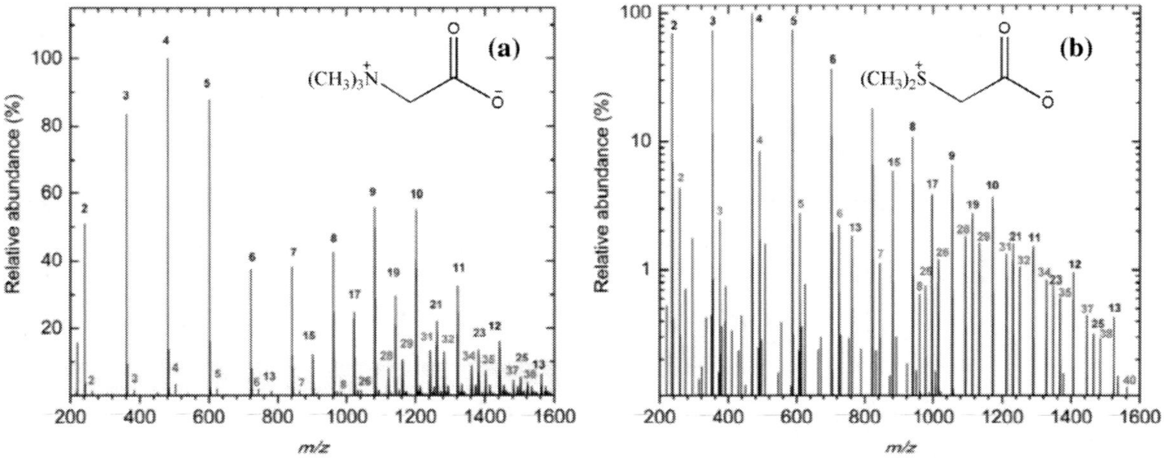

Figure 2. ESI high resolution MS of a solution of: (a) GB (Reproduced by permission of The Royal Society of Chemistry. [16]); (b) DMSA (Reprinted with permission from [12]. Copyright 2011 American Chemical Society). Both spectra contain series of singly charged clusters protonated $[M_n+H]^+$ (n=black) and sodiated $[M_n+Na]^+$ (n=green), doubly charged clusters $[M_n+2H]^{2+}$ (n=blue) and triply charged clusters $[M_n+3H]^{3+}$ (n=red).

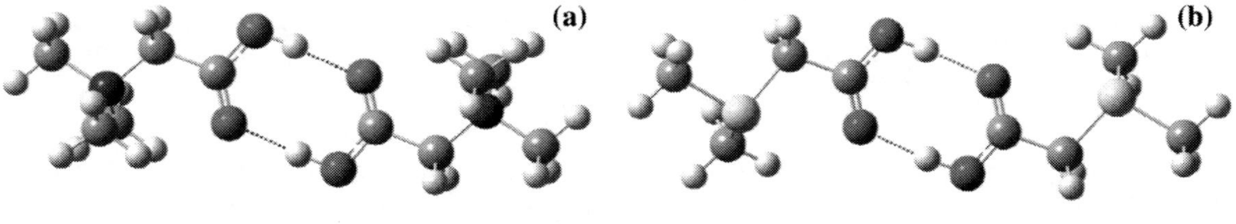

Figure 3. B3LYP/6-311++G(d,p) optimised structures of the $[M_2+2H]^{2+}$: (a) GB; (b) DMSA. Gaussian-03E01 [21] and Gaussian-09A02 program packages [22] were used.

Figure 3 shows the structures of the optimized doubly charged dimers $[M_2+2H]^{2+}$ for GB and DMSA. An examination of the dissociation of these dimers into singly charged monomers showed that both doubly charged dimers are thermodynamically unstable, but are kinetically stable. The dissociation of these dimers into two singly protonated monomers $[M+H]^+$ is predicted to be exothermic by -1.12 eV for the GB and -1.06 eV for the DMSA dimer, but is associated with an activation barrier of 0.20 eV for the GB and 0.22 eV for the DMSA dimer. The energy released in the dissociation of 1.12 eV for the GB and 1.06 for the DMSA doubly charged dimer corresponds to a charge separation of 12.9 Å and 13.6 Å, respectively, as calculated using Coulomb energy of two point charges in vacuum separated by distance, d. These values are in accordance with the sizes of the dimers, where the furthest two H are 13.1 Å apart in the case of the GB dimer and 13.6 Å in the case of the DMSA dimer. These results clearly demonstrate that it is the Coulomb repulsion of the two charges, which drives the dissociation of the doubly charged dimer.

3.2. Electron-induced fragmentation reactions of doubly charged zwitterionic betaines of GB $[M_{15}+2H]^{2+}$, $[M_{21}+2H]^{2+}$ and DMSA $[M_{15}+2H]^{2+}$

The mass spectra arising from the interactions of electrons with the $[M_{15}+2H]^{2+}$ cluster of GB as a function of electron energies ranging from 1.8 - 26.8 eV are shown in figure 4a-d. The ion-electron interactions of the $[M_{21}+2H]^{2+}$ cluster of GB as a function of electron energy from 2.1 – 32.1 eV are shown in figure 4e-h. The ion-electron interactions recorded for the DMSA cluster $[M_{15}+2H]^{2+}$ are shown in figure 5a-d. The investigation of the fragmentation of these clusters as a function of electron energy showed four types of fragmentation reactions discussed in the following sections.

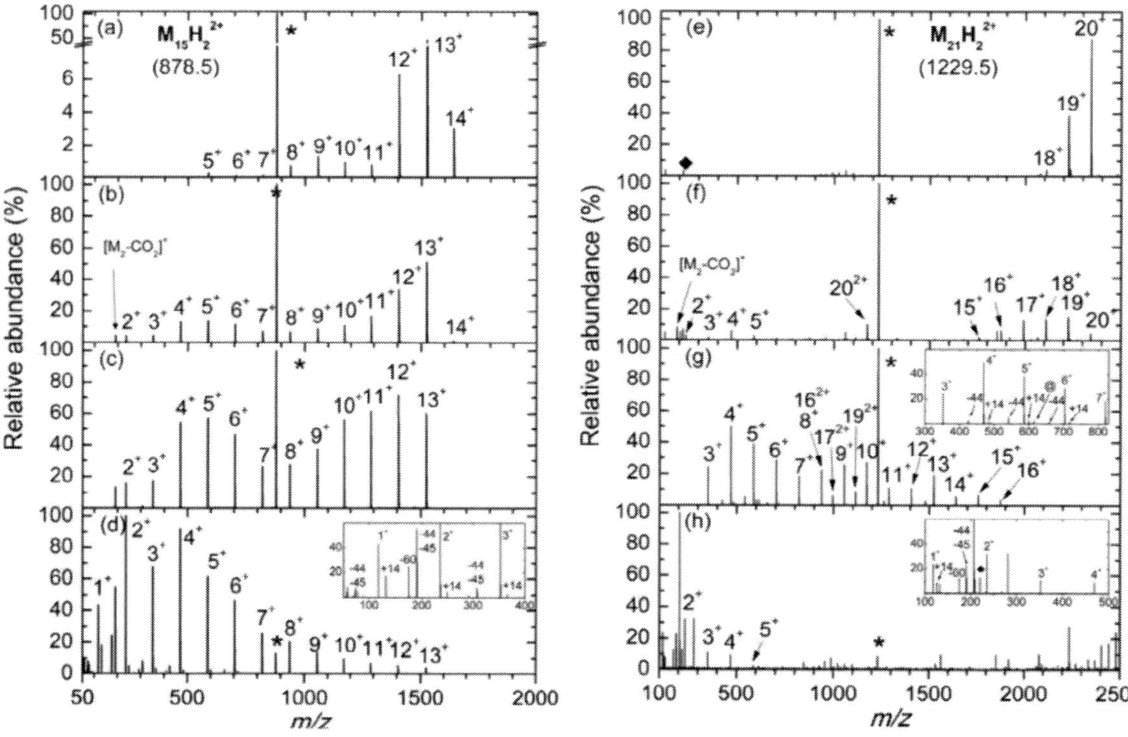

Figure 4. Ion-electron interactions of GB clusters: (left) the $[M_{15}+2H]^{2+}$ cluster at electron energies of: (a) 1.8 eV; (b) 11.8 eV; (c) 21.8 eV; and (d) 26.8 eV (Reprinted with permission from [12]. Copyright 2011 American Chemical Society); (right) the $[M_{21}+2H]^{2+}$ cluster at electron energies of (e) 2.1 eV; (f) 12.1 eV; (g) 17.1 eV; and (h) 32.1 eV (Adapted from reference [11].). All unassigned peaks are due to the background. In the inset of the figures, the numbers -44, -45, -60 and +14 stand for the mass difference of this fragment and the closest singly charged fragment $[M_n+H]^+$ and represent the loss of CO_2, CO_2H, $[C_2H_4O_2]$ and CH_3 group transfer, respectively. A * represents the mass-selected precursor ion. The ion designated by @ corresponds to the second harmonic peak in the FT-ICR and the ion designated by ◆ corresponds to background electronic noise.

3.2.1. *Electron capture.* At low-electron energies up to 12 eV mass spectra are dominated by large singly protonated fragment clusters, which suggests that the interaction of the doubly charged cluster $[M_n+2H]^{2+}$ with slow electron is dominated by electron capture, forming the charged reduced species $\{[M_n+2H]^{\bullet+}\}^*$ in an excited state, which stabilizes through the ejection of an H atom and at least one or more neutral molecules to form observed singly charged fragments $[M_n+H]^+$ of n=(n-1),(n-2),(n-3), etc. (equation (1), figure 4a, 4b and figure 5a).

$$[M_n+2H]^{2+} + e^- \rightarrow \{[M_n+2H]^{\bullet+}\}^* \rightarrow [M_{n-m}+H]^+ + M_m + H^\bullet \quad (1)$$

3.2.2. *Coulomb explosion.* Coulomb explosion of the charge is observed at electron energies from 12 eV in an asymmetric fission. Coulomb explosion proceeds either through direct excitation of the parent $\{[M_n+2H]^{2+}\}^*$ or through a singly charged transient intermediate $\{[M_n+2H]^{\bullet+}\}^*$ followed by electron detachment. The latter however, would be possible only if the states of the excited singly charged intermediate couple to the vibrational continuum of the excited doubly charged parent [23]. The higher electron energy of 26.8 eV (figure 4d, 4h and figure 5c, 5d) provides substantial excitation of the cluster, leading to a distribution of singly charged clusters mainly at lower mass, which likely comes from neutral evaporation of the singly charged clusters formed through Coulomb explosion. In the case of the $[M_{15}+2H]^{2+}$ cluster of GB, Coulomb explosion leads to a symmetric distribution of the fragment clusters on both sides of the parent peak (figure 4b and 4c), which showed the fragmentation to be dominated by charge separation. All possible singly charged fragments are observed. Coulomb explosion of the $[M_{21}+2H]^{2+}$ clusters shows highly asymmetric fission with singly charged products carrying away more than 70% of parent mass (figure 4f and equation (2)). Related asymmetric Coulomb explosion reactions have been observed in the water clusters, $[(H_2O)_n+2H]^{2+}$ [24].

$$[M_n+2H]^{2+} + e^- \rightarrow [M_{n-m}+H]^+ + [M_m+H]^+ + e^- \quad (2)$$

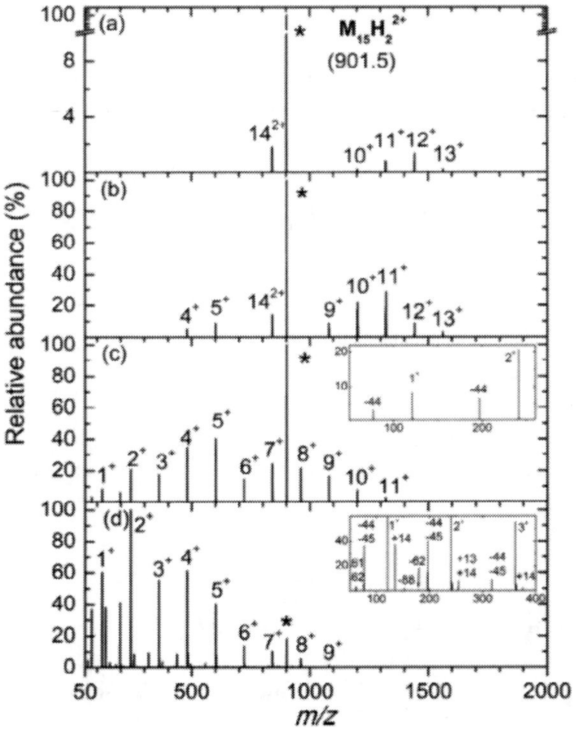

Figure 5. Ion-electron interactions of the $[M_{15}+2H]^{2+}$ DMSA cluster at electron energies of: (a) 1.8 eV; (b) 11.8 eV; (c) 21.8 eV; and (d) 26.8 eV. All unassigned peaks are due to the background. In the inset of the figures, the numbers -44, -45, -62, -88, +14 and +13 stand for the mass difference of this fragment and the closest singly charged fragment $[M_n+H]^+$ and represent the loss of CO_2 , CO_2H, $(CH_3)_2S$, $(CO_2)_2$ and CH_3 or CH_2 group transfer, respectively. A * represents the mass-selected precursor ion. (Reprinted with permission from [12]. Copyright 2011 American Chemical Society)

In the case of $[M_{15}+2H]^{2+}$ DMSA cluster, the Coulomb explosion observed at 11.8 eV through the detection of the tetramer $[M_4+H]^+$, pentamer $[M_5+H]^+$ coupled with the abundance of $[M_{11}+H]^+$ and

$[M_{10}+H]^+$ showed again an asymmetric fission, and perhaps a formation of preferred size domains (figure 5b). Note, that the tetramer $[M_4+H]^+$ and pentamer $[M_5+H]^+$ clusters in the ESI spectrum (figure 2b) have high abundance, whereas the intensity of $[M_6+H]^+$ abruptly drops by more than 50 %. Similarly, the intensity of $[M_9+H]^+$ and $[M_{10}+H]^+$ is higher than that of singly charged clusters on either side of the m/z scale. The distribution of singly charged clusters of DMSA is not as uniform as in the case of GB (figure 2a).

3.2.3. *Neutral evaporation.* Neutral evaporation was observed only in the case of DMSA for the $[M_{15}+2H]^{2+}$ cluster close to the stability limit (figure 5a, 5b) and in the case of the large GB $[M_{21}+2H]^{2+}$ cluster (figure 4f, 4g). The $[M_{15}+2H]^{2+}$ cluster dissociates through neutral evaporation forming $[M_{14}+2H]^{2+}$ at low electron energies up to 11.8 eV, which contrasts to GB clusters of the same size. This result supports the observation that the binding energy of the neutral DMSA in the cluster is lower than in the case of GB [12]. The $[M_{21}+2H]^{2+}$ cluster of GB undergoes neutral evaporation observed in the electron energy region between 12 and 25 eV (equation (3), figure 4f and 4g).

$$[M_n+2H]^{2+} + e^- \rightarrow [M_{n-m}+2H]^{2+} + M_m + e^- \quad (3)$$

3.2.4. *Bond cleavage reactions.* The first fragment due to the decarboxylation appeared at 11.8 eV as $[M_2\text{-}CO_2]^+$ in the EID of both GB $[M_n+2H]^{2+}$ clusters (n=15,21), however, such a type of fragments is most abundant at higher electron energies with a number of singly protonated fragment clusters that have lost a CO_2, or CO_2H (see inset of the figure 4d, 4g and of the figure 5d).Chemistry observed within the cluster resulted in fragment ions that acquired a CH_3 group from another molecule within the cluster prior the dissociation. This CH_3 group transfer has been observed for all clusters of betaines. The loss of 60 mass units shown in the inset of figure 4d corresponds to loss of $[C_2,H_4,O_2]$ that has been reported previously in the EID of the singly protonated GB dimer [25]. The bond cleavage reactions and intercluster chemistry induced by high energy electrons with clusters of DMSA lead to a higher variety of fragments. Thus, additional losses of $(CO_2)_2$ and $(CH_3)_2S$ were observed, as well as fragment clusters having an extra CH_2 group.

3.3. *Summary on electron interactions with positively multiply charged clusters*

Results of the investigation of the electron interactions with positively multiply charged clusters of betaines showed:

- Electrons are captured at low electron energies < 10 eV; for clusters close to the stability limit, electron capture competes either with Coulomb explosion, as in the case of GB, or neutral evaporation, as in case of DMSA, depending on the binding interaction of the neutrals within the cluster.
- At electron energies > 12 eV Coulomb explosion of the cluster competes with neutral evaporation; this competition is stronger for large clusters than clusters close to the stability limit.
- Coulomb explosion proceeds via asymmetric fission.
- The higher the electron energy, the more fragments are observed due to the bond cleavages. These fragments formed through bond cleavages of the cluster constituents seem to depend on the energy of electrons rather than the size of the cluster, which could bare an implication towards understanding chemistry induced in the solution.

3.4. Negatively charged clusters of tryptophan

ESI of a solution of Trp yields a series of cluster ions of the type $[Trp_n\text{-}xH]^{x-}$, x=1-4 (figure 6). The source conditions played a role in the relative abundances of the cluster ions observed, especially for higher charge states. Some of these clusters showed to be metastable, since they were observed to survive in the linear ion trap but did not make it through to the FT-ICR mass analyser [13]. Interestingly, the signal for singly charged clusters diminishes as n increases and the high resolution data revealed that the signal from the apparent $[Trp_4\text{-}H]^-$ is largely due to $[Trp_8\text{-}2H]^{2-}$. The smallest dianionic cluster observed in both, the ESI (observed in the linear ion trap) and via evaporation of Trp in CID of $[Trp_9\text{-}2H]^{2-}$ (see [13]) was $[Trp_7\text{-}2H]^{2-}$. The smallest triply charged cluster observed was $[Trp_{16}\text{-}3H]^{3-}$, and the smallest quadruply charged cluster appeared to be $[Trp_{31}\text{-}4H]^{4-}$. The difference in the "appearance size" n_a between adenosine-monophosphate clusters [26] and Trp clusters formed via ESI may be due to the difference in the size of the system (molecular weight, geometric structure) in relation to the Rayleigh limit [27,28] and/or to the existence of stronger acidic sites on the AMP cluster. Interestingly, the appearance size of Trp clusters, plotted in a log-log plot as a function of charge state q, appears to be a straight line (figure 7), a phenomenon previously observed with some atomic and molecular clusters [26,32].

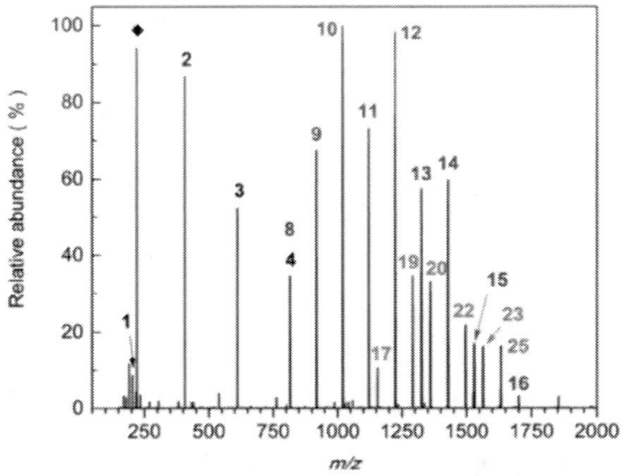

Figure 6. ESI high resolution MS of a solution of Trp. Spectrum contains series of singly charged clusters deprotonated $[M_n\text{-}H]^-$ (n=black), doubly charged clusters $[M_n\text{-}2H]^{2-}$ (n=blue) and triply charged clusters $[M_n\text{-}3H]^{3-}$ (n=red). The ion designated by ◆ corresponds to background electronic noise. (Adapted from reference [13].)

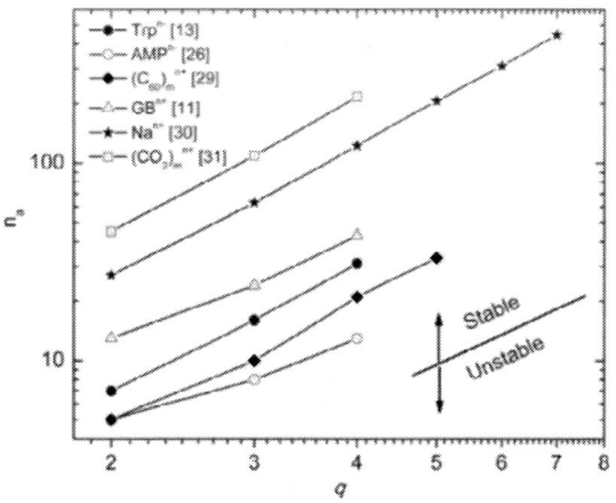

Figure 7. Log-log plot of the appearance size n_a of some cationic and anionic atomic and molecular clusters as a function of the charge q.

The EID spectra of $[Trp_9\text{-}2H]^{2-}$ clusters are shown in figure 8a-c (for all spectra measured see [13]). Investigation of the EID of $[Trp_9\text{-}2H]^{2-}$ cluster anions showed the following dissociation reactions: (i) neutral evaporation at low electron energies up to 21.8 eV (equation 4); the smallest doubly charged fragment observed was in this case also $[Trp_7\text{-}2H]^{2-}$; (ii) Coulomb explosion from 19.8 eV of electron energy; (iii) ejection of an electron at low electron energy up to 13.8 eV through observation of the formation of small yields of the singly charged clusters $[Trp_x\text{-}H]^-$ (x=3-8) (figure 8a). It seems that not all of these ions are arising from charge separation reactions (equation (5)) as not all have their counter ions. Thus, other processes must be involved. It was suggested that ejection of an electron occurs from the doubly deprotonated cluster with concomitant excitation of the resultant cluster radical anion, which then ejects the neutral radicals $[Trp\text{-}H]^{\bullet}$ or $[Trp_2\text{-}H]^{\bullet}$ to yield the observed spectrum (figure 8a); (iv) bond cleavages, observed in the electron interactions of positively charged cluster, were in this case not as pronounced. At 21.8 eV (see [13]), a small amount of $[Trp_2\text{-}H\text{-}NH_3]^-$ fragment ions was detected. A very intriguing observation was the abrupt loss of the parent and fragment ion signal in a particular electron energy region. In the electron energy region of 16.8-19.8 eV, no fragments were observed, and at 18.8 eV not even the $[Trp_9\text{-}2H]^{2-}$ precursor ion (see figure 9). Two possible explanations were suggested: (i) double electron loss from each of the $[Trp\text{-}H]^-$ carboxylates; (ii) ionization of two neutral Trp molecules to form neutral salt-bridge clusters of the type $\{[Trp]^{\bullet+}[Trp\text{-}H]^-[Trp]^{\bullet+}[Trp\text{-}H]^-,(Trp_5)]\}$, since the ionization energy of neutral Trp is 8.43 eV [33].

$$[Trp_9\text{-}2H]^{2-} \rightarrow [Trp_{9\text{-}y}\text{-}2H]^{2-} + yTrp \quad (4)$$
$$\rightarrow [Trp_{9\text{-}x}\text{-}H]^{-} + [Trp_x\text{-}H]^{-} \quad (5)$$

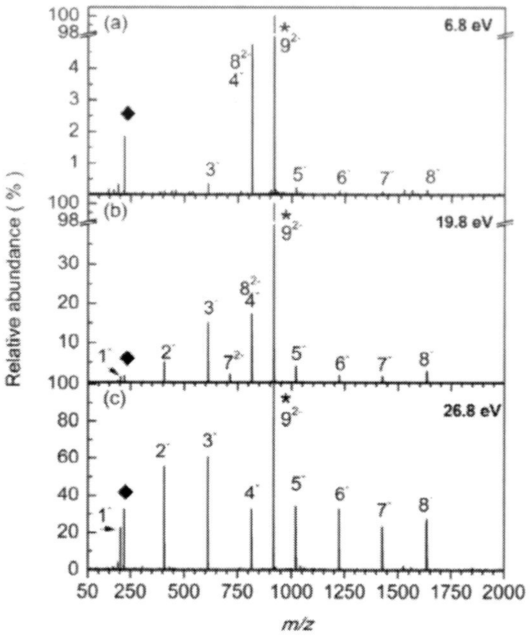

Figure 8. Ion-electron interactions of the $[Trp_9\text{-}2H]^{2-}$ cluster at electron energies of: (a) 6.8 eV; (b) 19.8 eV; (c) 26.8 eV. The ion designated by ◆ corresponds to the background electronic noise in the FT-ICR. All unassigned peaks are due to background noise. A * represents the mass selected precursor ion. (Adapted from reference [13].)

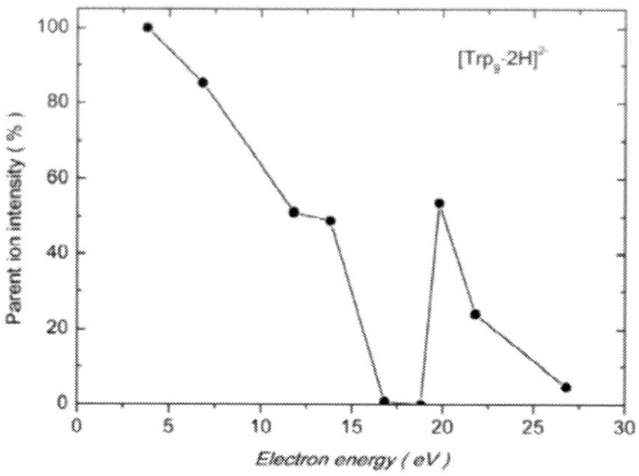

Figure 9. Parent ion intensity of the $[Trp_9\text{-}2H]^{2-}$ cluster as a function of electron energy.

3.5. *Summary on electron interactions with negatively multiply charged clusters*

The EID spectra of $[Trp_9\text{-}2H]^{2-}$ and the types of fragment ions observed are sensitive to the energy of the electrons, which is entirely consistent with the EID reactions of the clusters of betaine (section 3.2.). The electron interactions lead to:

- Competition of neutral evaporation and reactions leading to singly charged fragments that are not due to the Coulomb explosion at low electron energies.
- Competition of neutral evaporation and Coulomb explosion at electron energies > 19 eV.
- One minor fragment due to the bond cleavage of the cluster constituent at 21.8 eV.

4. Conclusions

Electron interactions were investigated (i) at different electron energies ranging from > 0 – 50 eV; (ii) with different cluster sizes above the stability limit; and (iii) with different charge states of biomolecular clusters. The following conclusions can be drawn:

- Electron interactions with positively and negatively doubly charged clusters do depend on the energy of the electrons.

- For both, positively and negatively charged clusters, neutral evaporation is in competition with reactions giving rise to singly charged clusters, i.e. Coulomb explosion and/or charge reduction via redox reactions.
- Coulomb explosion proceeds via asymmetric fission.

Fragmentation reactions of these multiply charged clusters were investigated as a function of the electron energy for the first time and the results showed a high potential to study these due to the excellent high mass resolution FT-ICR technique equipped with an electron cathode. Because the fragmentation reactions and the chemistry induced by electrons depend on the energy of impinging electrons, the capacity of this technique should be more explored with other biomolecular systems to help to understand the consequences of radiation damage.

Acknowledgements
LF thanks the ARC for the award of an APD Fellowship and for financial support via the ARC Centre of Excellence for Free Radical Chemistry and Biotechnology. The author gratefully acknowledges contribution of: EJ-H Yoo, GN Khairallah and RAJ O'Hair (Melbourne); TD Märk and P Scheier (Innsbruck); SB Nielsen (Aarhus). LF acknowledges the generous allocation of computing time from the Victorian Partnership for Advanced Computing (VPAC) Facility. An ARC Lief grant and funding from the Victorian Institute for Chemical Sciences are acknowledged for the purchase of the LTQ-FT.

References

[1] von Sonntag C 1987 *The chemical basis of radiation biology* (Taylor and Francis, London)
von Sonntag C 1991 *Physical and Chemical Mechanisms in Molecular Radiation Biology* (Plenum Press, New York)
[2] Uehara U, Nikjoo H and Goodhead D T 1999 *Radiat. Res.* **152** 202
LaVerne J A and Pimblott S M 1995 *J. Phys. Chem.* **99** 10540
[3] Scifoni E, Surdutovich E, Solov'yov A, Pshenichnov I, Mishustin I and Greiner W 2008 *AIP Conf. Proc.* **1080** 104
Surdutovich E, Obolensky O I, Scifoni E, Pshenichnov I, Mishustin I, Solov'yov A V and Greiner W 2009 *Eur. Phys. J.* D **51** 63
[4] Boudaiffa B, Cloutier P, Hunting D, Huels M A and Sanche L 2000 *Science* **287** 1658
Sanche L 2005 *Eur. Phys. J.* D **35** 367
Denifl S, Zappa F, Mahr I, Lecointre J, Probst M, Märk T D and Scheier P 2006 *Phys. Rev. Lett.* **97** 043201
Ptasinska S, Denifl S, Gohlke S, Scheier P, Illenberger E and Märk T D 2006 *Angew. Chem. Int. Ed.* **45** 1893
[5] Zappa F, Denifl S, Mähr I, Lecointre J, Rondino F, Echt O, Märk T D and Scheier P 2007 *Eur. Phys. J.* D **43** 117
Ferreira da Silva F, Denifl S, Märk T D, Ellis A M and Scheier P 2010 *J. Chem. Phys.* **132** 214306
[6] Schlathölter T, Alvarado F, Bari S, Lecointre A, Hoekstra R, Bernigaud V, Manil B, Rangama J and Huber B A 2006 *ChemPhysChem* **7** 2339
Huber B A, Adoui L, Bernigaud V, Manil B, Maunoury L, Rangama J, Rousseau P, Haag N, Johansson H, Schmidt H T, Cederquist H, Nielsen S B, Liu B, Zettergren H, Hvelplund P, Alvarado F, Bari S, Hoekstra R, Postma J and Schlathölter T 2008 *AIP Conf. Proc.* **1080** 21
[7] Yamashita M and Fenn J B 1984 *J. Phys. Chem.* **88** 4671
Fenn J B, Mann M, Meng C K, Wong S F and Whitehouse C M 1989 *Science* **246** 64
Cole R B 1997 *Electrospray Ionization Mass Spectrometry: Fundamentals, Instrumentation, and Applications* (John Wiley: New York)
[8] Benesch J L P, Ruotolo B T, Simmons D A and Robinson C V 2007 *Chem. Rev.* **107** 3544
Cooks R G, Zhang D, Koch K J, Gozzo F C and Eberlin M N 2001 *Anal. Chem.* **73** 3646
Nanita S C and Cooks R G 2006 *Angew. Chem. Int. Ed.* **45** 554

Takats Z, Nanita S C, Cooks R G, Schlosser G and Vékey K 2003 *Anal. Chem.* **75** 1514
[9] Rahman O A and Grotemeyer J 2005 *Eur. J. Mass Spectrom.* **11** 295
Marksteiner M, Haslinger P, Sclafani M, Ulbricht H and Arndt M 2009 *J. Phys. Chem.* A **113** 9952
[10] Haselmann K F, Jorgensen T J D, Budnik B A, Jensen F and Zubarev R A 2002 *Rapid Commun. Mass Spectrom.* **16** 2260
Prell J S, O'Brien J T, Holm A I S, Leib R D, Donald W A and Williams E R 2008 *J. Am. Chem. Soc.* **130** 12680
[11] Feketeová L and O'Hair R A J 2009 *Rapid Commun. Mass Spectrom.* **23** 3259
[12] Yoo E J-H, Feketeová L, Khairallah G N and O'Hair R A J 2011 *J. Phys. Chem.* A **115** 4179
[13] Feketeová L, Khairallah G N, Brunet C, Lemoine J, Antoine R, Dugourd P and O'Hair R A J 2010 *Rapid Commun. Mass Spectrom.* **24** 3255
[14] More Malek R, Metelmann-Strupat W, Zeller M and Muenster H 2005 *Am. Biotech. Lab.* **23** 8
Horning S, Malek R, Wieghaus A, Senko M W and Syka J E P 2003 *Proc. 51st ASMS Conf. Mass Spectrometry and Allied Topics* Montreal
[15] Feketeová L, Khairallah G N and O'Hair R A J 2008 *Eur. J. Mass Spectrom.* **14** 107
[16] Feketeová L and O'Hair R A J 2008 *Chem. Commun.* 4942
[17] Middlemiss D S, Facchini M, Morrison C A and Wilson C 2007 *CrystEngComm* **9** 777
Gora R W, Grabowski S J and Leszczynski J 2005 *J. Phys. Chem.* A **109** 6397
The core of the triply charged clusters may be a cyclic trimer. See: Roy A K and Thakkar A J 2004 *Chem. Phys. Lett.* **386** 162
[18] Gronert S 1999 *J. Mass Spectrom.* **34** 787
[19] Lee S-W and Beauchamp J L 1999 *J. Am. Soc. Mass Spectrom.* **10** 347
Kass S R 2005 *J. Am. Chem. Soc.* **127** 13098
[20] Harmon K M, Avci G F, Madeira S L, Mounts P A and Thiel A C 2001 *J. Mol. Struct.* **597** 177
[21] Gaussian 03, Revision E.01: Frisch M J et al. 2004 *Gaussian, Inc., Wallingford CT*
[22] Gaussian 09, Revision A.02: Frisch M J et al. 2009 *Gaussian, Inc., Wallingford CT*
[23] Smith I W M 1980 *Kinetics and dynamics of elementary gas reactions* (Butterworths: London, Boston)
[24] Gotts N G, Lethbridge P G and Stace AJ 1992 *J. Chem. Phys.* **96** 408
[25] Yoo E J-H, Feketeová L, Khairallah G N, White J M and O'Hair R A J 2011 *Org. Biomol. Chem.* **9** 2751
[26] Liu B, Hvelplund P and Nielsen S B 2005 *J. Am. Soc. Mass Spectrom.* **16** 1840
[27] Rayleigh L 1882 *Philos. Mag.* **14** 185
[28] Echt O, Scheier P and Märk T D 2002 *C. R. Physique* **3** 353
[29] Manil B, Maunoury L, Huber B A, Jensen J, Schmidt H T, Zettergren H, Cederquist H, Tomita S and Hvelplund P 2003 *Phys. Rev. Lett.* **91** 215504-1
[30] Näher U, Frank S, Malinowski N, Zimmermann U and Martin T P 1994 *Z. Phys. D* **31** 191
[31] Echt O, Sattler K and Reknagel E 1982 *Phys. Lett.* **90A** 185
[32] Concina B, Hvelplund P, Nielsen A B, Nielsen S B, Rangama J, Liu B and Tomita S. 2006 *J. Am. Soc. Mass Spectrom.* **17** 275
[33] Wang X B, Yang X, Nicholas J B and Wang L S 2001 *Science* **294** 1322

Nano-IBCT 2011 IOP Publishing
Journal of Physics: Conference Series **373** (2012) 012010 doi:10.1088/1742-6596/373/1/012010

Electron interactions with tetrahydrofuran

M.C. Fuss[1], R. Colmenares[2], A.G. Sanz[1], A. Muñoz[3], J.C. Oller[3], F. Blanco[4], T.P.T. Do[5], M.J. Brunger[5], D. Almeida[6], P. Limão-Vieira[6], and G. García[1]

[1] Instituto de Física Fundamental, Consejo Superior de Investigaciones Científicas, Serrano 113-bis, 28006 Madrid
[2] Servicio de Oncología Radioterápica, Sección Radiofísica, Hospital Universitario Ramón y Cajal, crta. Colmenar Viejo km 7.1, 28034 Madrid, Spain
[3] Centro de Investigaciones Energéticas, Medioambientales y Tecnológicas (CIEMAT), Avenida Complutense 22, 28040 Madrid, Spain
[4] Departamento de Física Atómica, Molecular y Nuclear, Universidad Complutense de Madrid, 28040 Madrid
[5] ARC Centre for Antimatter-Matter Studies, School of Chemical and Physical Sciences, Flinders University, G.P.O. Box 2100, Adelaide, SA 5001, Australia
[6] Atomic and Molecular Collisions Laboratory, Department of Physics, New University of Lisbon, 2829-516 Caparica, Portugal

g.garcia@iff.csic.es

Abstract. In this paper, we summarize our recent experimental and theoretical results on electron scattering from gaseous tetrahydrofuran (THF). Electron-impact ionization and total scattering cross sections were determined experimentally for energies between 50-5000 eV. Electron energy loss spectra were measured in the keV range using a transmission beam technique and for smaller energies (15-50 eV) with a crossed-beam apparatus. Using an optical potential method assuming the screening-corrected additivity rule, total, elastic and inelastic cross sections including dipole interactions were calculated (1eV - 10keV) in order to complement the experimental data. Elastic differential cross sections were also obtained. An empirical approximation to the inelastic angular distributions based on differential cross sections is proposed. The available integral and differential cross sections and energy loss distributions in the range 1 eV - 10 keV are combined into a table of recommended electron interaction cross sections with THF.

1. Introduction

It is well known [1] that high energy radiation produces abundant secondary electrons ($\sim 4 \cdot 10^4$ per MeV of energy deposited), which are the main source of the energy transfer map and radiation damage in biological tissues. Even electrons with sub-ionisation energies can produce damage, in terms of DNA strand breaks and molecular dissociation, more efficiently than the traditionally considered mechanism of direct ionisation of the medium (Sanche and colleagues [2]). Unfortunately, DNA is not itself readily amenable for the studies needed to determine a complete set of electron interaction data, for example when it comes to modelling radiation effects in a simulation. As a consequence, moieties of DNA like tetrahydrofuran (THF), as well as water, can serve as model systems for attempting to build up the requisite data bases for track simulations in matter [3].

Here we summarize recent measurements and calculations by our group on different aspects of electron scattering from THF and present new energy loss distributions and elastic differential cross sections. In section 2, we compare the experimental and theoretical total cross sections (CS). Section 3 deals with ionization cross sections. Elastic differential cross sections are presented in section 4. Finally, in section 5 electron energy loss (EEL) spectra for two different energy ranges are presented. Based on these distributions, we derive electronic excitation cross sections and propose an empirical approximation for inelastic angular distributions in the low energy range. In the discussion and conclusion section, we aim to establish a set of recommended electron-THF cross section data and present these in resulting tables.

2. Total scattering cross sections

Absolute total electron scattering cross sections for gaseous THF were obtained previously in the range 50eV – 5keV, by measuring the electron beam attenuation in a transmission beam experiment [3]. In brief, the apparatus employed consisted of an electron gun, an interaction chamber containing the target gas, and a hemispherical energy analyzer followed by an electron channel multiplier as a detector. Three electrostatic quadrupoles along the beam's path were used for focussing and aligning the electrons. The experimental uncertainty of the present measurement is estimated to be below 5%, according to the analysis of the main error sources carried out in [4].

The optical potential method described in previous papers [5,6] has been used to calculate differential and integral elastic, as well as integral inelastic, electron-THF scattering CS for energies up to 10 keV. This calculation was based on an independent atom model (IAM) but using a screening-corrected additivity rule (SCAR) and included recent adjustments [6,7] especially important in the low energy region. Processes involving nuclear motion were neglected in this calculation. The present method considered inelastic scattering as being due to electron-electron interaction processes; only those arising from electronic excitation were taken into account, so that vibrational excitations are ignored. However, rotational excitation was included by taking into account dipole scattering contributions.

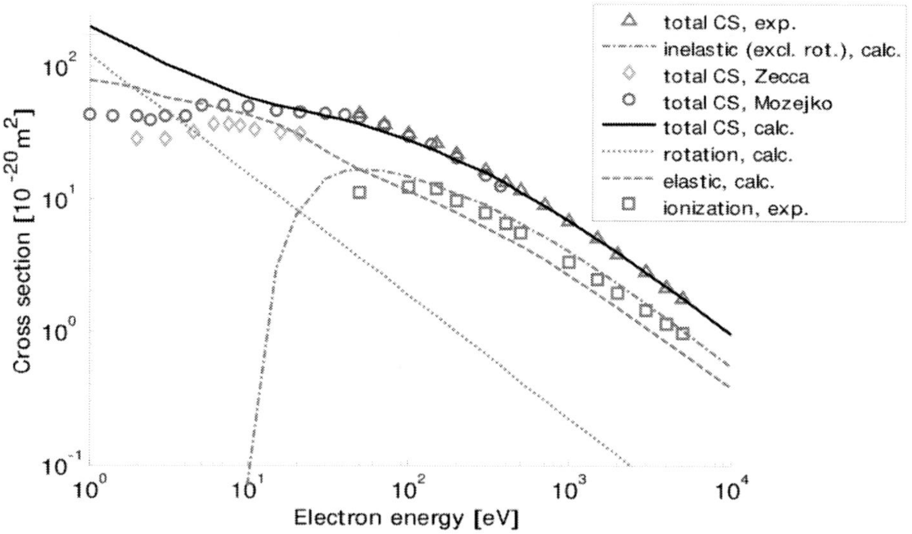

Figure 1. Total scattering cross sections, obtained experimentally in Madrid, compared to the results of our IAM-SCAR calculations. Experimental total cross sections published by Zecca et al. [9] and Mozejko et al. [8] et el. are also included.

The total scattering cross sections obtained both experimentally and theoretically [3] are plotted in figure 1. An excellent agreement between our experiment and calculation can be observed in the keV range. For lower energies, differences were found to be ≤ 10%. On the other hand, the present experimental results are within 7% of those presented by Mozejko et al. [8] for the overlapping energy range. For low energies (< 15 eV), we find a large difference between our calculation and experiments by other groups. This is not surprising since an independent atom model, even with screening corrections, cannot be expected to give accurate predictions for such low energies. On the other hand, the large dipole moment of the THF molecule enhances rotational excitation particularly for low energies (see figure 1). This is reflected in a higher theoretical total CS due to the fact that theory deals with an ideal system with all target molecules in their ground state. This is, however, not a realistic premise for an experimental system at room temperature where rotationally excited levels are populated.

3. Ionization

Electron-impact ionization was determined previously [3] for energies between 50 eV and 5 keV, by combining simultaneous electron and ion measurements with a time-of-flight analysis of the ionic fragments. In brief, electrons were generated by an emitting filament and then driven into a gas cell by a pulsed voltage. This collision chamber featured two perpendicular apertures, each leading to a drift tube. Positive ions and secondary electrons were simultaneously detected in the two opposite directions, by two-stage micro-channel plate assemblies. The beam current transmitted through the gas cell was also measured in order to normalize the ionization signal and thus obtain ionization cross sections. The total electron-impact ionization cross sections for THF [3] can be seen in figure 1. The associated uncertainty is estimated to be about 7% including uncertainties in the normalization procedure. We observe very good agreement of about 6% with the theoretical (binary-encounter-Bethe) data of [10].

4. Elastic differential cross sections

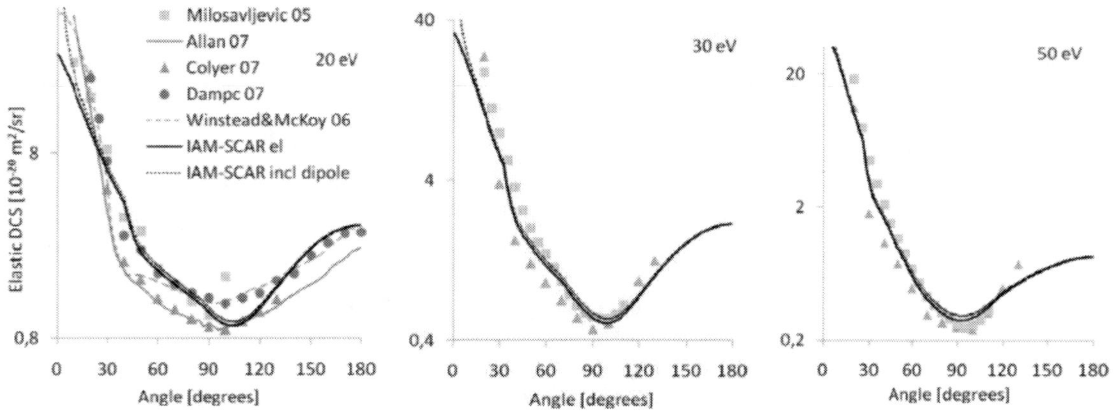

Figure 2. Comparison of our IAM-SCAR elastic DCS (black solid line) to results obtained in other experiments [12-15] and calculations [16].

In addition to integral scattering cross sections, elastic differential cross sections (DCS) were also calculated with the IAM-SCAR method. A comparison of our results with different experiments and the calculations of Winstead and McKoy [16] is displayed in figure 2. Similar to what has been observed for other molecules such as methane [11] and ethylene, these theoretical angular distributions agree well with the experimental data of Milosavljevic et al. [12] in the higher energy range. At 50 eV (their upper limit), we also find good agreement with the results of Colyer et al. [13] both in shape and magnitude. Below about 30 eV, increasing discrepancies become apparent between the experimental [13-15] and theoretical data. Our calculation does not predict the steep increase of the elastic cross section at forward scattering angles (below 30°-40°) which is present in the experimental results. Agreement does become better, however, when including dipole interactions in our calculation. A further difference lies in the localization of the global minimum of the angular distributions, which appears shifted towards higher angles in the theoretical DCS at energies < 30 eV.

5. Electron energy loss distributions, electronic excitation and inelastic DCSs

EEL distributions were measured for two incident electron energy ranges. From 300 eV up to 1000 eV, spectra were recorded in Madrid in the same apparatus described in section 2 and used for the total CS measurements. In this case for the energy loss measurements, the electrostatic quadrupoles were employed to tilt the beam's trajectory and thus obtain a mixture of small-angle scattered electrons at the analyzer entrance while avoiding the primary beam. In figure 3, the average distributions obtained for different energies are presented. As can be seen, variations between those spectra are minimal and do not follow any clear trend with energy. The average over all distributions (right panel of figure 3) yields a mean energy loss for the processes covered (electronic excitation, ionization, and neutral dissociation) of 26.7 eV. Note the excellent agreement, taking into account the poorer resolution, in the overlapping range (onset of the inelastic region) of this average distribution with the average of the 50 eV spectra recorded at Flinders University.

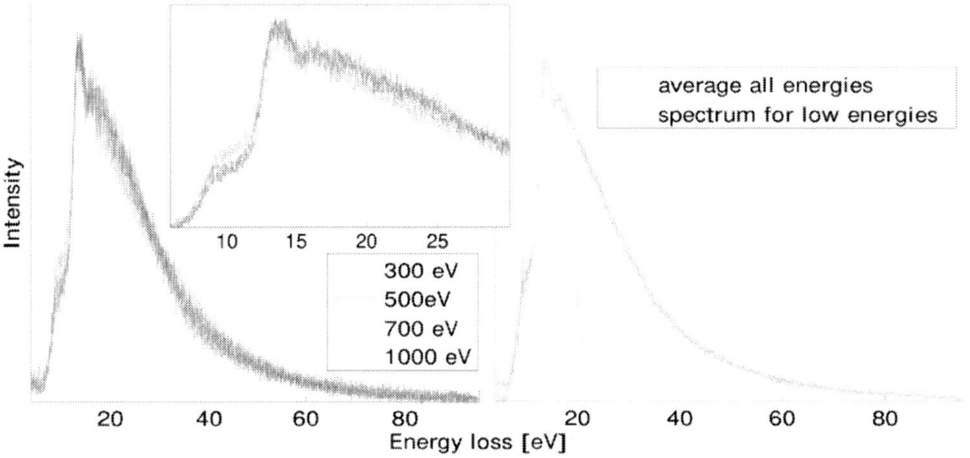

Figure 3. Energy loss spectra obtained for the incident electron energy range 300-1000 eV. The right plot shows the average of all distributions measured. The dashed line illustrates how it could be adapted for use with a lower incident electron energy (30 eV).

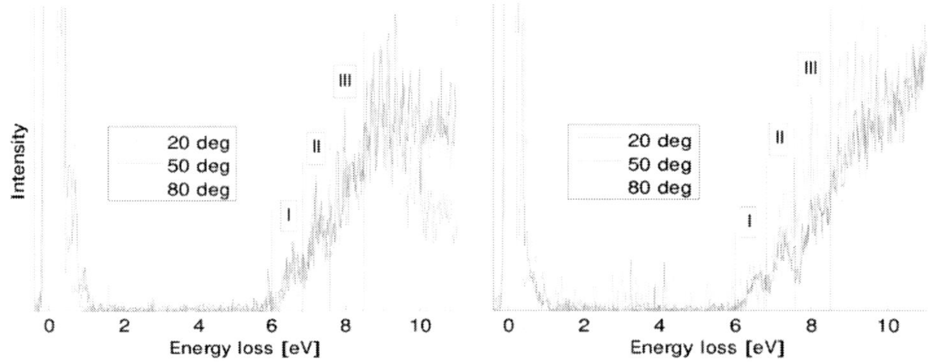

Figure 4. Example of the energy loss spectra obtained at Flinders University at 20 eV (left panel) and 50eV (right panel). Three energy bands I, II and III of Rydberg

states, as denoted in the figure, were analyzed separately.

For the incident energies 15 - 50 eV and energy losses up to 11 eV, spectra were recorded at Flinders University for well-defined angles from 15° to 90° [12]. The energy resolution achieved in those experiments was 50-60 meV. Figure 4 shows some spectra obtained at 20 eV and 50 eV. In order to derive differential cross sections for electronic excitation, the energy loss region was first divided in three bands of Rydberg states as denoted in the plot. A fitting procedure adapted to accommodate the particular spectroscopy of THF [14], was then used to yield the ratio of the DCS for the band of interest (band I, band II, or band III), to that for the elastic DCS [13]. In this way, differential electronic excitation cross sections were derived [12]. Integral electronic excitation cross sections were determined using a molecular phase shift analysis technique for extrapolation to 0° and 180°.

Since experimental inelastic differential cross sections are in many cases only available for a restricted range of scattering angles and incident energies, applications like electron transport simulations often have to rely on approximations in spite of the growing efforts to obtain accurate inelastic angular distributions. This is why the present differential energy loss spectra were further examined, in search of an empirical formula approximating inelastic scattering behaviour based on elastic DCS (generally more easily available).

When inspecting the experimental angular distributions of electrons scattered inelastically (after having excited the target molecule electronically), we observed that as a general trend, differences in magnitude (maxima and minima) become smaller for increasing energy loss ΔE and scattering into large angles becomes more important. This means that that the steep increase in the DCS at forward angles present for low energy losses, similar to the elastic DCS, is much less prominent for larger energy losses relative to the incident energy. An electron having excited a higher energy level or caused ionization, is thus more likely scattered by a larger angle than an electron having excited the lowest state. A diagram of this qualitative trend is displayed in figure 5.

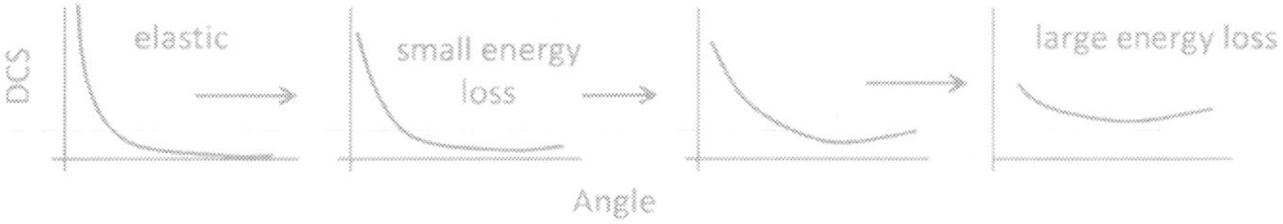

Figure 5. Schematic of the general behaviour observed for elastic and inelastic angular distributions as a function of the energy loss.

A suitable approximation should reproduce these findings. For reasons of mathematical consistency, we also require our distribution to tend towards the elastic DCS for $\Delta E \rightarrow 0$. Taking those points into account, we propose an approximation of the inelastic angular distributions by:

$$\frac{\partial^2 \sigma}{\partial\Omega \partial\Delta E} \propto \left(\frac{\partial\sigma}{\partial\Omega}\right)_{el}^{1-\frac{\Delta E}{E}}, \qquad (1)$$

for a given incident energy E, where σ is the cross section and Ω a solid angle. Absolute values (DCS) can be obtained by normalization to the corresponding integral CS. One can easily see that this approximation yields the elastic CS $(\partial\sigma/\partial\Omega)_{el}$ for ΔE=0 and produces an isotropic distribution for the extreme case of complete energy loss, ΔE=E. In fact it thus represents a gradual transition between two popular approximations, namely the supposition of isotropic scattering and the direct use of elastic DCS for inelastic events.

This formula is of course a very rough approximation to a complex problem encompassing several, physically diverse, collision processes. However, it is obviously more accurate than the two before-mentioned methods, while being easy to implement in any simulation and not requiring extensive calculations. The distributions resulting for 15 eV, 30 eV and 50 eV as well as the corresponding experimental points are displayed in figure 6. The main difference to the method considering momentum transfer, used previously for the cases of water and methane [20,11], is the relatively larger weight for the backscattering. When comparing the two methods, we noted an improvement in emulating the experimental points with the present approximation. A quantitative evaluation shows a mean deviation from the experimental data of 17% (present formula) vs. 50% ("momentum transfer" approximation) at 15 eV incident energy, 13% vs. 22% at 30 eV and 20% vs. 29% at 50 eV. The present approximation did also compare favourably to the previous method in the cases of methane and ethylene [21] and will shortly replace it in our simulation code LEPTS [22].

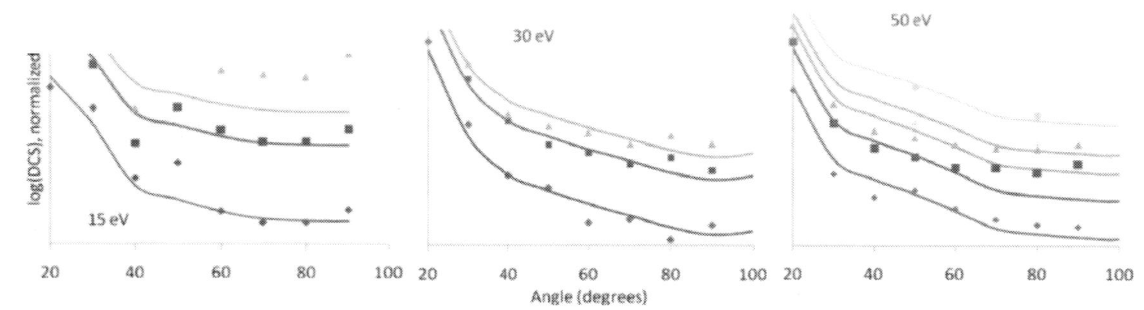

Figure 6. Inelastic angular distributions obtained with the present approximation. Experimental data for energy losses of: ◆, 6.45 eV; ■, 7.3 eV; ▲, 8.1 eV; +, 9.15 eV; ●, 10.5 eV.

6. Discussion and conclusion

Taking into account the results obtained by our group and complementary data taken from literature, a self-consistent set of recommended data can be put together for the energy range 1 eV - 10 keV. The general strategy followed here is to give priority to experiments and use theoretical data only if no adequate experimental values are available.

Regarding integral cross sections, the different interaction channels are studied separately and then their sum compared to the total CS, which contains the lowest uncertainty and is taken to be accurate. The recommended total scattering CS are the present total scattering CS completed with data from [8] for low energies. Present calculated elastic CS show best agreement with the experiment from Allan [14], which does furthermore cite the best energy resolution, suggesting a lower contamination from rotational excitation processes. Recommended elastic CS are thus based on those two sets of results. The vibrational excitation CS values represent an estimation based on the absolute integral values and differential excitation functions published in [14]. Since no experimental results are available for rotational excitation, values were taken from our calculation and adjusted in absolute magnitude to the remaining CS (σ_{tot}-σ_{el}-σ_{vib}) below the onset of ionization, electronic excitation and neutral dissociation. Electronic excitation CS are based on the Flinders measurements [12], already described, and their extrapolation to higher energies. These data might well fall somewhat below the true value of integral electronic excitation since levels above ~ 8.5 eV are ignored. However, one can argue that this might be approximately compensated for by the fact that some excitations lead to neutral dissociation which is counted separately in its own channel. Recommended ionization CS originate from our experiment and the calculation of Mozejko and Sanche [10]. The remaining CS for energies $\geq$ 15 eV (σ_{tot}-σ_{el}-σ_{vib}-σ_{rot}-σ_{ion}-σ_{exc}) are attributed to neutral dissociation not otherwise accounted for. The resulting interaction data set is therefore presented in table 1.

Due to the good agreement between our calculated elastic DCS and the experiments at higher energies, we recommend using IAM-SCAR results for $\geq$ 30 eV. For lower energies, experimental data by [13,14] should be used for any application. A selection of the resulting recommended elastic angular distributions, for the whole energy range, is presented in figure 7. The complete table for incident energies in the range 1-10000 eV and angles 0°-180° is available online as supplementary data. For obtaining inelastic angular distributions, the empirical formula proposed in section 5 can be used together with the recommended elastic DCS.

Table 1. Recommended integral cross sections in $10^{-20}\,m^2$.

E (eV)	total	elastic	rotation	vibration	electronic excitation	ionisation	neutral dissociation
1	43.4	20.4	21.75	1.22	-	-	-
1.5	42.6	26.2	15.25	1.16	-	-	-
2	42.7	29.6	11.80	1.29	-	-	-
3	42.2	32.2	8.25	1.76	-	-	-
4	42.8	33.6	6.40	2.30	-	-	-
5	50.7	41.0	5.20	4.54	-	-	-
7	51.3	41.7	3.86	5.74	-	-	-
10	49.8	41.4	2.80	5.6	-	-	-
15	46.5	37.2	1.94	3.28	1.997	1.63	0.456
20	46.2	31.1	1.50	1.94	5.056	4.38	2.18
30	44.4	23.0	1.03	0.901	5.563	8.37	5.54
40	43.2	19.2	0.795	0.523	5.502	10.6	6.55
50	41.0	16.8	0.645	0.343	4.849	11.2	7.16
70	36.5	13.9	0.475	0.182	1.977	12.0	8.02
100	31.4	11.5	0.340	0.093	0.948	12.5	6.02
150	26.3	9.27	0.235	0.043	0.411	12	4.34
200	22.0	7.90	0.180	0.0250	0.227	9.89	3.78
300	16.6	6.19	0.125	0.0116	0.0986	7.9	2.28
400	13.7	5.15	0.095	0.0067	0.0545	6.57	1.82
500	11.8	4.45	0.075	0.0044	0.0344	5.64	1.60
700	9.08	3.53	0.055	-	0.0172	4.37	1.10
1000	6.96	2.70	0.040	-	0.0083	3.39	0.822
1500	5.11	1.95	0.027	-	0.0036	2.49	0.639
2000	3.90	1.54	0.021	-	-	1.99	0.349
3000	2.84	1.09	0.015	-	-	1.46	0.276
4000	2.17	0.851	0.011	-	-	1.16	0.148
5000	1.78	0.700	0.009	-	-	0.978	0.093
7000	1.327	0.518	0.006	-	-	0.750	0.052
10000	0.958	0.378	0.005	--	--	0.568	0.007

Since no significant variation with energy has been observed in our high-energy EEL spectra, a unique, average distribution (figure 3, right panel) is recommended to be used for the whole range of energies studied here. For applications, particularly interested in low energies (≤50 eV), that distribution can be adapted by including the excitation detail of the low energy spectra and by forcing a faster decrease after the ionization maximum, as sketched in figure 3 by the dashed line for 30 eV.

Summarizing, here we have used previous and new experimental and theoretical results in order to establish a complete set of electron-THF interaction data in the range 1 eV - 10 keV. This data base includes integral and differential cross sections for elastic and inelastic interactions and an energy loss distribution. Among other applications, it can be used as input parameters for the Low Energy Particle Track Simulation (LEPTS) code [22] for simulating electron transport through THF.

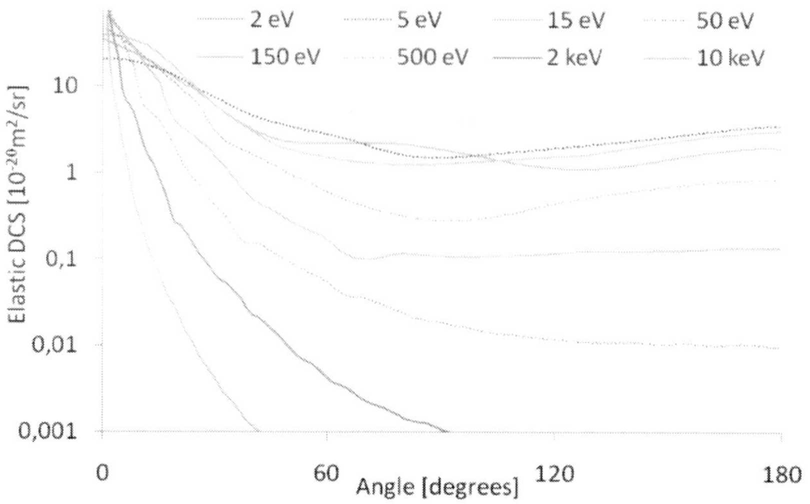

Figure 7. Recommended elastic DCS.

Acknowledgements
This work has been carried out with partial financial support of the Spanish Ministerio de Ciencia e Innovación (Project FIS2009-10245), the European Science Foundation (COST Actions CM0601 and MP1002) and the Australian Research Council's Centre of Excellence Program.

References

[1] Plante I and Cucinotta F A 2009 *New J. Phys.* **11** 063047
[2] Boudaiffa B, Cloutier P, Hunting D, Huels M A and Sanche L 2000 *Science* **287** 1658; Boudaiffa B, Cloutier P, Hunting D, Huels M A and Sanche L 2002 *Radiat. Res.* **157** 227; Huels M A, Boudaiffa B, Cloutier P, Hunting D and Sanche L 2003 *J. Am. Chem. Soc.* **125** 4467
[3] Fuss M, Muñoz A, Oller J C, Blanco F, Almeida D, Limão-Vieira P, Do T P T, Brunger M J and García 2009 *Phys. Rev.* A **80** 052709
[4] Williart A, Kendall P A, Blanco F, Tegeder P, García G and Mason N J 2003 *Chem. Phys. Lett.* **375** 39
[5] García G, Blanco F and Williart A 2001 *Chem. Phys. Lett.* **335** 22; Blanco F and García G 2003 *Phys. Rev.* A **67** 022701
[6] Blanco F and García G 2003 *Phys. Lett.* A **317** 458
[7] Blanco F and García G 2004 *Phys. Lett.* A **330** 230
[8] Mozejko P, Ptasinska-Denga E, Domaracka A and Szmytkowski C 2006 *Phys. Rev.* A **74** 012708
[9] Zecca A, Perazzolli C and Brunger M J 2005 *J. Phys.* B **38** 2079
[10] Mozejko P and Sanche L 2005 *Rad. Phys. Chem.* **73** 77.
[11] Fuss M C, Muñoz A, Oller J C, Blanco F, Hubin-Franskin M-J, Almeida D, Limão-Vieira P and García G 2010 *Chem. Phys. Lett.* **486** 110
[12] Milosavljević AR, Giuliani A, Šević D, Hubin-Franskin M-J and Marinković B P 2005 *Eur. Phys. J.* D **35** 411-416
[13] Colyer C J, Vizcaino V, Sullivan J P, Brunger M J and Buckman S J 2007 *New J. Phys.* **9** 41
[14] Allan M 2007 *J. Phys.* B **40** 3531–3544
[15] Dampc M, Milosavljević A R, Linert I, Marinković B P and Zubek M 2007 *Phys. Rev.* A **75** 042710

[16] Winstead C and McKoy V 2006 *J. Chem. Phys.* **125** 074302
[17] Do T P T, Leung M, Fuss M, Garcia G, Blanco F, Ratnavelu K and Brunger M J 2011 *J. Chem. Phys.* **134** 144302
[18] Brunger M J and Teubner P J O 1990 Phys. Rev. A **41** 1413
[19] Giuliani A, Limão-Vieira P, Duflot D, Milosavljević A R, Marinković B P, Hoffmann S V, Mason N, Delwiche J and Hubin-Franskin M-J 2009 *Eur. Phys. J.* D **51**, 97
Bremner L J, Curtis M G and Walker I C 1991 *J. Chem. Soc. Faraday Trans.* **87**, 1049
[20] Muñoz A, Oller J C, Blanco F, Gorfinkiel J D, Limão-Vieira P, Maira-Vidal A, Borge M J G, Tengblad O, Huerga C, Téllez M and García G 2008 *J. Phys. Conf. Ser.* **133** 012002
[21] Fuss M C and García G 2012, personal communication.
[22] McEachran R P, Sullivan J P, Buckman S J, Brunger M J, Fuss M C, Muñoz A, Blanco F, White R D, Petrović Z Lj, Limão-Vieira P and García G 2012 *J. Phys.* B **45** 045207

Photoelectron spectra: a tool of analysis of irradiation dynamics

M. Baer[1], **P. M. Dinh**[2], **P.-G. Reinhard**[1,2] **and E. Suraud**[1,2]
[1]Institut für Theoretische Physik, Universität Erlangen, D-91058 Erlangen, Germany
[2]Laboratoire de Physique Théorique, Université Paul Sabatier, CNRS, F-31062 Toulouse Cédex, France

E-mail: dinh@irsamc.ups-tlse.fr

Abstract. We explore from a theoretical perspective photoelectron spectra emitted from a Na_8 cluster irradiated by a short laser pulse. We use time-dependent density-functional theory (TDDFT) at the level of the local-density approximation (LDA). That includes a self-interaction correction (SIC) to recover the proper emission threshold. As environments are known to play an important role in irradiation dynamics, we consider the example of Na_8 deposited on a MgO(001) surface and compare it to the case of a free cluster. Anisotropy of emission is also briefly discussed.

1. Introduction

For decades, optical methods have provided a key tool for analyzing cluster structure and dynamics [1–4]. More recently, new insights into the dynamics of irradiation have been gained by measuring the properties of electrons emitted after irradiation by a short laser pulse. The first step in that direction is photoelectron spectroscopy (PES) which measures kinetic energies of emitted electrons. While early attempts in this direction were performed long ago [5, 6], steady development of tools allows now to measure fully angular and energy resolved distributions of outgoing electrons [7–10].

The case of clusters in contact with environments (embedded clusters or clusters deposited on a substrate) brings further challenging aspects into play and furthermore allows to understand the impact of an environment on emission properties, a key aspect in many physical situations ranging from materials science to biological systems. Contact with an environment may be compulsory in biological compounds and it is often used for easier experimental handling in molecular and cluster physics. On the other hand, it complicates the physical situation and its theoretical description. For example, deposition on a surface modifies emission properties and furthermore requires to take into account a huge number of degrees of freedom to accommodate an acceptable (possibly dynamical) description of the substrate. The theoretical description of such problems thus constitutes a demanding task. This paper aims at addressing such a question for a rather simple example with a clear-cut separation between optically active system and environment. We thus perform a first theoretical exploration of PES of laser excited metal clusters deposited on an insulating surface MgO(001).

The description is based on density-functional theory [11], which is well adapted for electronic structure and dynamics in clusters [12]. The detailed dynamics of laser excitation and subsequent electron emission is treated by the time-dependent extension of DFT, namely TDDFT [13,14]. As a test case, we take Na_8 deposited on a MgO(001) surface which is a typical insulator material. We concentrate on a moderate laser excitation in which electronic emission is dominated by direct emission [15–17]. In practice, there is competition between direct and thermal emission, the preferred exit channel depending on pulse duration. For sufficiently short pulses, collisional relaxation and the resulting thermal emission can be neglected. We will use pulses with full width at half maximum (FWHM) of 20 fs.

2. Outline of the model

We use a hierarchical model in which the cluster is described in full microscopic detail while the environment is treated at a lower level of refinement with classical molecular dynamics. This model corresponds to an improved QM/MM approach as frequently used in biochemistry or surface science [18]. The new aspect is that electronic excitations are possible at the side of the substrate by attributing to each substrate atom a dynamical polarization. The details of this model have been presented at many places and we refer the reader to the recent review [19] which summarizes many aspects of the model and applications. For completeness sake, we recall a few key aspects.

The constituents of the system and their degrees of freedom are: $\varphi_n(\vec{r},t)$, $n = 1, ..., N_{\rm el}$ for valence electrons of the Na cluster, $\vec{R}_{i^{(\rm Na)}}$, $i^{(\rm Na)} = 1, ..., N_{\rm i}$ for the positions of the Na^+ ions, $\vec{R}_{i^{(c)}}$, $i^{(c)} = 1, ..., M$ for the positions of the O cores, $\vec{R}_{i^{(v)}}$, $i^{(v)} = 1, ..., M$ for the centers of the O valence clouds, $\vec{R}_{i^{(k)}}$, $i^{(k)} = 1, ..., M$ for the positions of the Mg^{2+} cations. The Na cluster is treated in standard TDDFT at the level of local-density approximation, coupled with molecular dynamics, (TDLDA-MD) [12, 20]. The Mg^{2+} cations (label $i^{(k)}$) are electrically inert and treated as charged point particles. The polarization of O^{2-} anions is described by distinguishing between a valence electron distribution ($i^{(v)}$) and a complementing core ($i^{(c)}$). All ions of the MgO substrate are described as classical degrees of freedom in terms of positions $\vec{R}_{i^{(\rm type)}}$, the O^{2-} dipoles dynamically stemming from the difference $\mathbf{R}^{(c)} - \mathbf{R}^{(v)}$. The Mg and O ions are arranged in bulk MgO crystalline order (six layers of 784 Mg^{2+} plus 784 O^{2-} ions). The dynamical degrees of freedom for Mg and O are taken into account in an active cell of the MgO(001) surface region underneath the Na cluster (three layers, each with 242 Mg^{2+} plus 242 O^{2-}). The active cell is embedded in an "outer region" of MgO material with ions kept fixed, but oxygen dipoles still fully dynamical. Beyond that region, only the Madelung potential is considered. Finally the ions in the lowest layer are fixed to prevent relaxation towards an artificial second surface.

The total energy is composed as $E = E_{\rm Na} + E_{\rm Surf} + E_{\rm coupl}$ where $E_{\rm Na}$ describes an isolated Na cluster, $E_{\rm Surf}$ the MgO(001) substrate, and $E_{\rm coupl}$ the coupling between the two subsystems. For $E_{\rm Na}$, we take the standard TDLDA-MD functional [12, 20] including an average-density self-interaction correction [21]. The energy of the substrate and the coupling to the Na cluster consists of the long-range Coulomb energy and some short-range repulsion modeled through effective local core-potentials [22–24]. The model parameters were furthermore calibrated to represent results of calculations to a fully quantum-mechanical description of the active MgO cell [25].

The equations of motion are obtained by variation of the above energy in standard manner [12] and are augmented by an external laser field of frequency $\omega_{\rm las}$, intensity I with $\sin^2$ pulse profile activated only in the time interval $0 \leq t \leq T_{\rm pulse}$ which corresponds to a full width at half (intensity) maximum as FWHM$\approx T_{\rm pulse}/3$. The laser polarization is perpendicular to the surface (z-axis). The TDDFT equations are solved on a grid in coordinate space in full three dimensions [20].

3. Results

Experimentally, it is possible to measure the kinetic energy of the electrons which were ionized from the cluster by a laser pulse. For not too strong laser pulses, the resulting spectra allow to directly analyze the single-particle structure of the cluster. The theoretical calculations in [17, 26], which used the same technique as employed in this work, have confirmed that, at least for free clusters and not too intense pulses, the simple single-particle picture is valid. This is demonstrated in figure 1 (upper panel) for the case of a free Na_8 cluster irradiated by a laser with frequency $\omega = 0.18\,\mathrm{Ry}$, which is slightly below the plasmon frequency of the cluster.

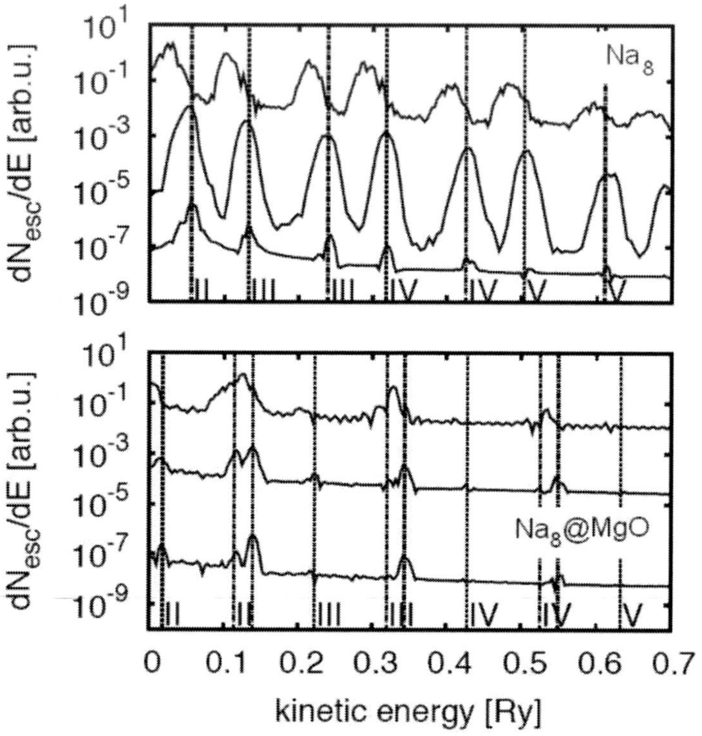

Figure 1. Kinetic energy distribution of emitted electrons from free Na_8 (top) and Na_8 on MgO(001) (bottom). The laser frequency is $\omega = 0.18\,\mathrm{Ry}$ (top) and $\omega = 0.205\,\mathrm{Ry}$ (bottom). The three curves correspond to laser intensities 10^8, 10^9 and $10^{10}\,\mathrm{W/cm^2}$ (same ordering from bottom to top in each panel) and have been shifted up/down for better visualization. The Roman numbers denote the multi-photon order responsible for the peak and the vertical lines display the peak position expected from the single-particle spectrum, identifying the orbitals from which the electron density was emitted (dotted line for $1s$, dashed-dotted for $1p$).

Figure 1 shows results from calculations using three different laser intensities, $I = 10^8, 10^9$ and $10^{10}\mathrm{W/cm^2}$. The photon energy is smaller than the ionization potential of the cluster, so that one photon does not suffice to lift up an electron into the continuum. Two- or more-photon processes are required. At moderate intensities, total ionization remains small, and the residual cluster does not charge up significantly, leaving the ground-state mean-field practically unchanged. In this case, peaks can be found in the spectrum at

$$E_{i,n} = \varepsilon_i + n \cdot \omega > 0 \quad , \tag{1}$$

where ε_i are the single-particle energies of the occupied Kohn-Sham orbitals of the ground state, and n is the multi-photon order. These energies are plotted as vertical lines in figure 1. The Roman numbers indicate the multi-photon order n. Obviously, for the two lower laser intensities, as mentioned above, the simple single-particle approach correctly describes all peaks in the spectra. The identification can further be confirmed by watching the motion of the peaks with changing ω: Changing ω by $\Delta\omega$ moves the peak $E_{i,n}$ by $\Delta E_{i,n} = n \cdot \Delta\omega$. For higher intensities (upper curve in figure 1) peaks are broadened and the whole spectrum is red-shifted, due to changing the mean-field (cluster charging). The reason is the changed mean-field in this case. The strong laser pulse ionizes considerably more electrons (about 0.7 in this case), charging the residual cluster, so that the remaining electrons are deeper bound. Going over to even higher intensities would completely wipe out the single-particle structure in the spectrum [26].

The same technique is now applied for the deposited cluster but with a laser frequency of 0.205 Ry instead. The first main question here is whether the single-particle picture is still valid. Figure 1 (bottom) indicates that this is indeed the case. All dominant peaks in the spectra can be clearly identified with multi-photon transitions from the ground-state single-particle Kohn-Sham orbitals. Note that the degeneracy of the $1p$ level is removed by the substrate (two close-by vertical lines instead of one for free Na_8) and that this $1p$ splitting is also found in the PES. The second question in focus is what happens when the outgoing electron wave hits the substrate and is reflected. The repulsive interface potentials prevent the electron wave from penetrating the surface, but it could possibly exchange energy with the surface dipoles. It is thus conceivable that the electronic scattering at the substrate is not elastic. If that was the case, one would expect that the peaks in the kinetic energy spectrum would broaden or satellite peaks would show up. However, in all computed cases, no peak broadening nor further dominant peaks, which cannot be explained by the single-particle structure or excitations spectrum, were found. So the inelasticity of the scattering process seems to be small. This is not so surprising as both the laser and the electronic frequencies involved reside in a frequency domain where the dielectric response of MgO is very small. Above $\sim 0.6\,\text{Ry}$, however, the dielectric function of MgO possesses a broad resonance, but the contributions of the outgoing electron wave in this energy region are quite small. Our model qualitatively reproduces the appearance of the resonance in the dielectric response (triggered by the oxygen shell mass), but not the details. So the model is probably not very reliable if the laser frequency is well above 0.6 Ry.

In the case of the deposited cluster, the kinetic energy distribution is expected to strongly depend on the direction of emission, i.e. on the location of the measuring points used to calculate the energy spectra. The distributions in figure 1 have been obtained by averaging over all measuring points. Figure 2 shows the spectra as calculated at three individual measuring points, one far above the surface on the positive z-axis ($\theta = 0$, later called A), another one $5\,a_0$ above the surface on the x-axis ($\theta = \pi/2$, called B), and finally one deeply below the surface on the z-axis ($\theta = \pi$, called C). Note that, since the substrate is modeled as a finite slab, the measuring point is located below the slab in empty space. The spectrum measured at A is very similar to that measured at B, apart from the fact that the relative weight of emission between the split $1p$ orbitals changes slightly. Anyway, the n-photon processes mapping the single-particle spectrum still dominate the distribution. The electron yield decreases with increasing photon order n as intuitively expected. This is clearly different for the spectrum measured at C: the electron yield does not decrease with increasing n. For instance, the figure shows that a 3-photon process from a $1p$ state ($E_{\text{kin}} = 0.26\,\text{Ry}$ in the figure) is two orders of magnitude more probable than a 2-photon process from the same $1p$ state ($E_{\text{kin}} = 0.08\,\text{Ry}$). This result is due to the insulating properties of the MgO substrate. The electrons do not gain enough energy by absorbing two photons in order to override the transmission barrier (at 0.125 Ry in figure 2), whereas absorption of 3 photons suffices because the emitted electrons have now acquired sufficient kinetic energy.

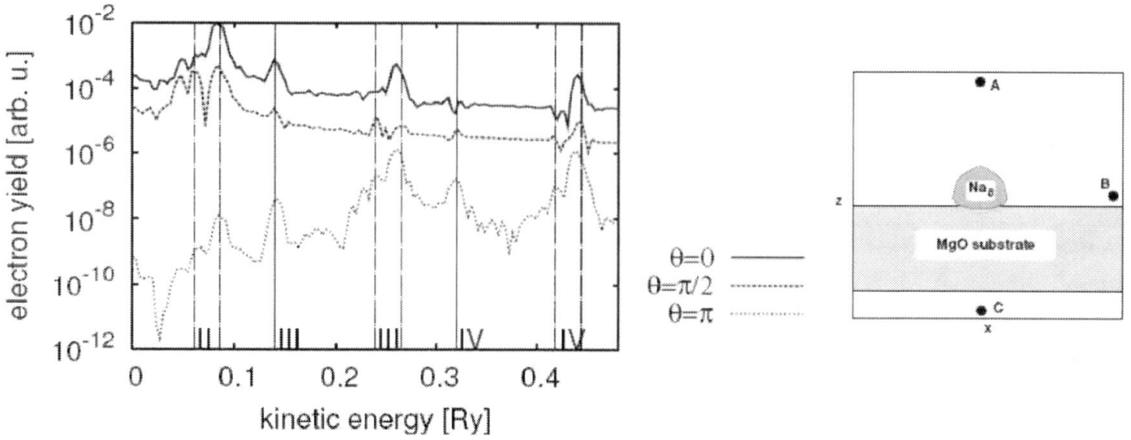

Figure 2. Kinetic energy distribution of emitted electrons from Na_8 exposed to a laser pulse with $\omega = 0.18\,\mathrm{Ry}$ as calculated from three different measuring points (see text for detail). The spectrum corresponding to $\theta = 0$ has been shifted up by a factor of 10 for better visualization.

Vice versa, the plot indirectly shows that the inelasticity in the scattering process between electron and MgO surface is only small. The electron density which has gained enough energy from the laser to penetrate the material, can pass the whole substrate slab and leave it on the lower side, where the point C resides. Even after passing the MgO layers, the principal kinetic energy distribution of the electron density is still dominated by the original electronic structure of the cluster, so that the electron has obviously not coupled to the oxygen dipoles. This, however, is not surprising, as the dielectric response of MgO is small in this energy/frequency domain. Resonances become important only for larger frequencies, see the discussion above.

4. Conclusions and perspectives

We have presented photoelectron spectra (PES) of a small sodium cluster deposited on an insulator MgO(001) surface and compared the properties of emitted electrons to the case of a free cluster. This serves as a simple archetype example for the analysis of the effects of an environment on PES. In both cases, there remains a relevant map of the single-particle energies of occupied states. Nonetheless, the surface significantly affects the energies of emitted electrons. That is to a large extent due to a shift of the single-particle energies emerging by deposition on a surface. The effect is even larger when considering angular distributions of emitted electrons, as suggested by comparing emission properties at various points in space. This is due to the fact that transmission of free electrons into the substrate is strongly suppressed by the large band gap of MgO(001). A next step is obviously to consider a full description of angular distributions of emitted electrons [27]. Another interesting aspect is to study PES in the case of irradiated organic molecules and to study the impact of the environment (for example water) in such cases. Work in this direction is in progress.

Acknowledgments

This work was supported, by Agence Nationale de la Recherche (ANR-06-BLAN-0319-02), the Deutsche Forschungsgemeinschaft (RE 322/10-1) and the Humboldt foundation.

References

[1] Brack M 1993 *Rev. Mod. Phys.* **65** 677
[2] de Heer W 1993 *Rev. Mod. Phys.* **65** 611
[3] Kreibig U and Vollmer M 1995 *Optical Properties of Metal Clusters, Springer Series in Materials Science 25* (Berlin: Springer)
[4] Haberland H 1994 *Clusters of Atoms and Molecules 1- Theory, Experiment, and Clusters of Atoms, Springer Series in Chemical Physics 52* (Berlin: Springer)
[5] McHugh K M *et al* 1989 *J. Chem. Phys.* **91** 3792
[6] Lichtenberger D L 1991 *et al* Chem. Phys. Lett. **176** 203
[7] Pinaré J C 1999 *et al Eur. Phys. J.* D **9** 21
[8] Baguenard B *et al* 2001 *Phys. Rev.* A **63** 023204
[9] Verlet J R R *et al* 2004 *J. Chem. Phys.* **121** 10015
[10] Kostko O *et al* 2007 *J. Phys. Conf. Ser.* **88** 012034
[11] Dreizler R M and Gross E K U 1990 *Density Functional Theory: An Approach to the Quantum Many-Body Problem* (Berlin: Springer)
[12] Reinhard P G and Suraud E 2003 *Introduction to Cluster Dynamics* (New York: Wiley)
[13] Runge E andGross E K U 1984 *Phys. Rev. Lett.* **52** 997
[14] Marques M A *et al* 2006 *Time Dependent Density Functional Theory, Lecture Notes in Physics 706* (Berlin: Springer)
[15] Campbell E E B *et al* 2000 *Phys. Rev. Lett.* **84** 2128
[16] Schlipper R *et al* 2001 *Appl. Phys.* A **72** 255
[17] Pohl A *et al* 2004 *J. Phys.* B **37** 3301
[18] Altoe P *et al* 2007 *Theor. Chem. Acc.* **118** 219
[19] Dinh P M *et al* 2010 *Phys. Rep.* **485** 43
[20] Calvayrac F *et al* 2000 *Phys. Rep.* **337** 493
[21] Legrand C *et al* 2002 *J. Phys.* B **35** 1115
[22] Nasluzov A *et al* 2001 *J. Chem. Phys.*, **115** 8157
[23] Reza Ahmadi G *et al* 1995 *Chem. Phys.* **199** 33
[24] Duplàa F and Spiegelmann F 1996 *J. Chem. Phys.* **105** 1492
[25] Bär M *et al* 2007 *Eur. Phys. J.* D **45** 507
[26] Pohl A *et al* 2000 *Phys. Rev. Lett.* **84** 5090
[27] Wopperer P *et al* 2010 *Phys. Rev.* A **82** 063416

Laser-plasma accelerator based femtosecond high-energy radiation chemistry and biology

Y A Gauduel

Laboratoire d'Optique Appliquée, CNRS UMR7639, Ecole Polytechnique ParisTech - ENS Techniques Avancées ParisTech, 91761 Palaiseau Cedex (France) and Radiation Biology MELUSYN Network, Armir-Collège de Polytechnique, Paris (France)

E-mail: yann.gauduel@ensta-paristech.fr

Abstract Regarding the different protocols used for external cancer radiotherapy (X or γ rays, electron, proton or ions beams) or radioimmunotherapy (Auger electron emitting radionuclides), the initial energy deposited in integrated biological systems (biomolecular and sub-cellular targets) represents a decisive parameter for the primary and more delayed radiation damage. A short-range energy distribution governs mainly [i)] the early survival probability of secondary electrons, [ii)] the spatio-temporal distribution of short-lived reactive radicals inside nascent tracks, [iii)] the primary biomolecular alterations triggered by low energy secondary electrons. The thorough understanding of these fundamental processes requires a real-time investigation of primary radiation events, typically in the temporal range 10^{-14} - 10^{-11} s. Laser-plasma accelerators based High Energy Radiation Femtochemistry (HERF) represents a newly emerging interdisciplinary field which can be driven in strong synergy with the generation of ultrashort particle beams in the MeV energy domain. The innovating developments of HERF would favour the investigation of prethermal radiation processes in aqueous and biochemically relevant environments. In this way, the quantum character of a very-short lived low-energy electron state (p-like configuration) represents a promising sub-nanometric probe to explore early radiation processes in native tracks. The specific properties of ultra-short electron beams accelerated by TW laser are very useful for future developments of spatio-temporal radiation biophysics in complex biological systems such as living cells.

1. Introduction

The innovative aspects of radiation therapy are growing rapidly as a result of new technical solutions enabling improved radiation sources, diagnostics and precision treatments. The exposure of living matter to ionizing radiations (X- and γ-rays, electrons, protons or ions) in the keV and MeV domains induces a broad range of complex physical responses and signalling processes that cells and tissues integrate or remove to maintain their functional integrity and prevent tumour formation. It is commonly admitted that the early spatial distribution of energy deposition triggered by ionizing radiation interactions with bio-molecular architectures is decisive for the control of damage at cellular or tissue levels and for the prediction of delayed responses in radiation medicine and cancer therapy [1-4]. The complex links that exist between the physical aspects of early radiation events and the delayed evolution of biological endpoints, carcinogenesis or cell survivals need the development of advanced *spatio-temporal radiation biomedicine* [5]. One major challenge of radiation biomedicine is to determine the fate of confined clusters of ionization and their associated radical processes in the nanometric environment of biological targets. In this way, the investigation of elementary radical reactions in condensed phase can be carried out in strong synergy with the recent progresses of ultra-short laser sources [6-10]. The innovating TW laser plasma accelerators that deliver ultra-short high energy electron bunches in the MeV range open promising perspectives for the development of high energy radiation femtochemistry (HERF) [11]. Considering the energy dependence of the electron stopping power in water for instance (figure 1), the real-time investigation of early radiation events in native tracks becomes accessible. This approach requires also the contribution of LERF (low energy radiation femtochemistry) devoted to the ultrafast spectroscopy of low energy radiation processes, typically for E < 6 eV.

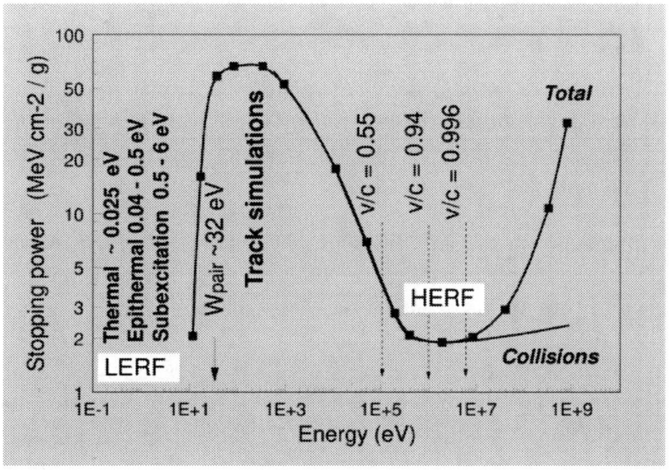

Figure 1. Energy dependence of electron stopping power in liquid water. The specific domains of energy covered by Low and High Energy Radiation Femtochemistry (LERF and HERF respectively) are reported on the figure.

2. Importance of a multiparametric approach including energy, space and time

2.1 Energy deposition and spatial distribution of early ionisation processes

The interactions of relativistic MeV electrons with water molecules induce ultrafast energy scattering processes and fractionated ionization clusters (figures 2 and 3). These ionization processes involve a hierarchy of electron populations for which the energy varies from the relativistic levels to the thermal value (kT ~0.025 eV). Typically, in less than 10^{-16}s, energy quanta of 200 and 20 eV are delivered in nanometric tracks and spurs, respectively. Due to the uncertainty relation for time and energy, the fastest ionising events in confined clusters take place in less than 0.33×10^{-16} s (equation 1).

$$\Delta t.\Delta E = \hbar = 6.6\ 10^{-16}\ \text{eV.s} \tag{1}$$

A second uncertainty principle for the position and momentum of relativistic particles must also be considered: $\Delta p.\Delta x = \hbar$ with p the momentum of the particle and $\Delta p = \Delta E/u$ if u represents the particle velocity (~3×10^{10} cm s^{-1}). From these two incertitude principles, an expression of Δx can be extracted (equation 2).

$$\Delta x = \hbar\,.\,{}^{u}\!/_{\Delta E} \tag{2}$$

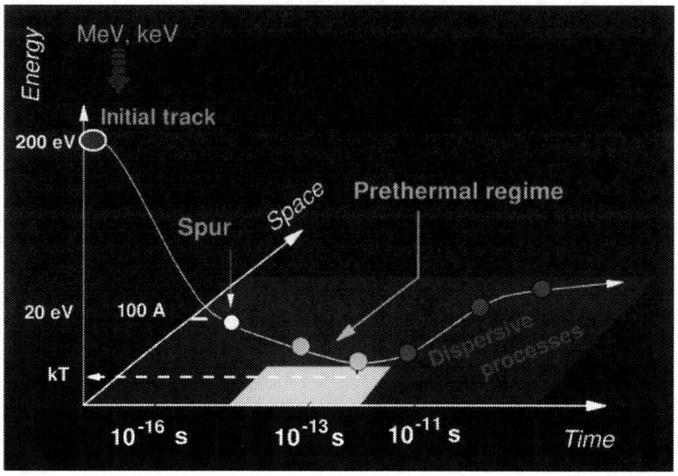

Figure 2. Time-space relationship of primary events in a water environment following an energy deposition triggered by MeV ionising radiation. In nanometric ionisation tracks, the prethermal regime of very low energy electrons corresponds to the temporal window 10^{-14}-10^{-12} s.

Consequently, a quantum energy of 20 eV deposited by relativistic particles occurs on 10^{-6} cm (~100 A) and involves nanometric ionization clusters. As early radiation damage can be highly dependent on the survival probability of low-energy secondary electrons and of the spatial distribution of primary radicals produced from water molecules, a thorough knowledge of native tracks requires the real-time probing of radiation events in the 10^{-15} – 10^{-11} s range. For an aqueous environment, this temporal domain concerns mainly prethermal events for which quantum states of very low excited electrons involve nonadiabatic transitions and lead to the sub-picosecond localization of secondary electrons in the water bath. Beyond 10^{-11} s, fully relaxed electrons contribute to sub-micrometric dispersive diffusion processes which can be carefully described by classical approaches such as the master diffusion equation [12,13].

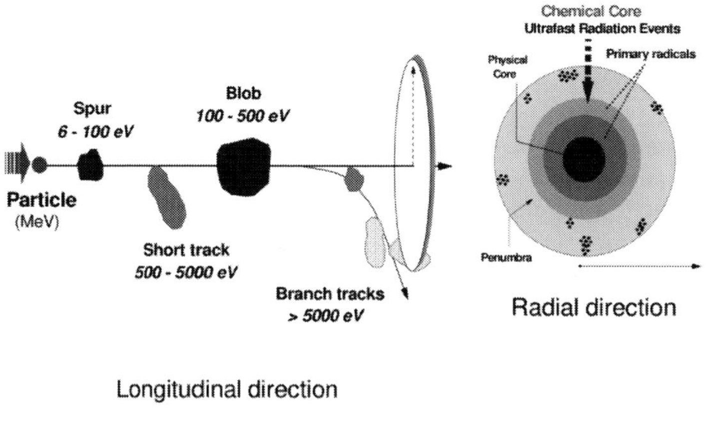

Figure 3. Schematic representation of a spatial distribution of inhomogeneous ionisation tracks following the interaction of relativistic MeV particles with an aqueous environment. In the radial direction of the radiation beam, early radiation events take place within the chemical core i.e. at the interface of the physical core and the penumbra zone.

2.2 Ultrafast irradiations by femtosecond high-energy electron bunches

The powerful laser techniques (table-top terawatt Ti:Sa laser amplifier systems) combined to laser plasma interactions provide femtosecond high-energy electron beams, typically in the 2.5 - 15 MeV range, which might conjecture the direct observation of primary radiation events in spurs and tracks. From the technical point of view, a 820 nm TW titanium-doped sapphire laser beam with an on-target energy of 960 mJ in 30 fs FWHM pulses can be focused onto the sharp edge of a 2 mm diameter supersonic helium gas jet. With intensities of the order of 2.7×10^{19} W/cm^2 in vacuum, the femtosecond laser pulse ionises the gas and excites relativistic plasma waves in the underdense plasma [7,10,14,15].The electrons of the plasma are accelerated by the intense electric field in the forward direction producing a well collimated electron beam with an initial duration of the same order of the TW laser pulse. Typically, a total charge of the electron beam determined by an integrating current transformer equals 2.5 ± 0.2 nC and corresponds to a mean number of $(1.55 \pm 0.15)\times10^{10}$ electrons. The specific qualities and properties of ultra-short relativistic particle bunches foreshadow the development of advanced research to deepen the biophysical understanding of ultrafast radiation events in native ionization tracks. Recent femtolysis experiments (Femto$_{\text{second radio}}$lysis) of aqueous targets, performed with pulsed electron bunches of 2.5-15 MeV were devoted to the investigation of early radical processes [16]. For an effective temperature of the femtosecond electron beam of about 4.5 MeV, the integrated energy flux in aqueous samples equals 5.8×10^{10} MeV cm^{-2}. Using a highly-time-resolved pump (femtosecond electron bunch) – optical probe (femtosecond photon pulse) orthogonal configuration, the real-time investigation of primary radiation processes can be performed in the radial direction of femtosecond electron bunches (figure 3).

Laser-plasma accelerator based high energy radiation femtochemistry gives new insights into the time dependence of fully hydrated electrons (e^-_{aq}) within native ionisation clusters. As expressed in equation 3 and shown in figure 4 the short-time dynamics of this ubiquitous radical (1s ground state) is mainly governed by its transient recombination with surrounding water products (hydronium ion and radical hydroxyl).

$$e^-_{aq}(t) = \int_{-\infty}^{t} \frac{dn_{e^-_{aq}}(t')}{dt} \left\{ 1-\gamma + \gamma \,\mathrm{erf}\left(\frac{T_J}{t-t'} \right)^{1/2} \right\} dt' \tag{3}$$

$$\text{with} \quad \mathrm{erf}(x) = 2/\sqrt{\pi} \int_0^u e^{-u^2} du$$

In this analytical expression, two adjustable parameters are considered: 1D geminate recombination time which is equivalent to a jump time (T_J) and a recombination probability γ. These random 1D walking electrons satisfy both the experimental absorption signal dynamics in the 0-75 ps range and the time-dependence $1/(t)^{1/2}$ of dispersive recombination processes inside non-homogeneous ionization clusters (multi-scale phenomena). The least-squares fit of the 820 nm signal dynamics emphasizes that dispersive geminate recombinations between e^-_{aq} and neoformed OH radical or H_3O^+ ion take place immediately after a femtosecond energy deposition in the water bath. Such non synchronized events are mainly observed between 0 and 20 ps. This early electron recombination dynamics is significantly influenced by the spatial distribution of ionisation events. Indeed, the recombination time (T_J) is much shorter in diluted ionization environments triggered by a two-photon excitation process (2×4 eV) than in confined native tracks induced by relativistic electron bunches: T_J = 1.2 ps vs 2.0 ps, respectively (upper part of figure 4).

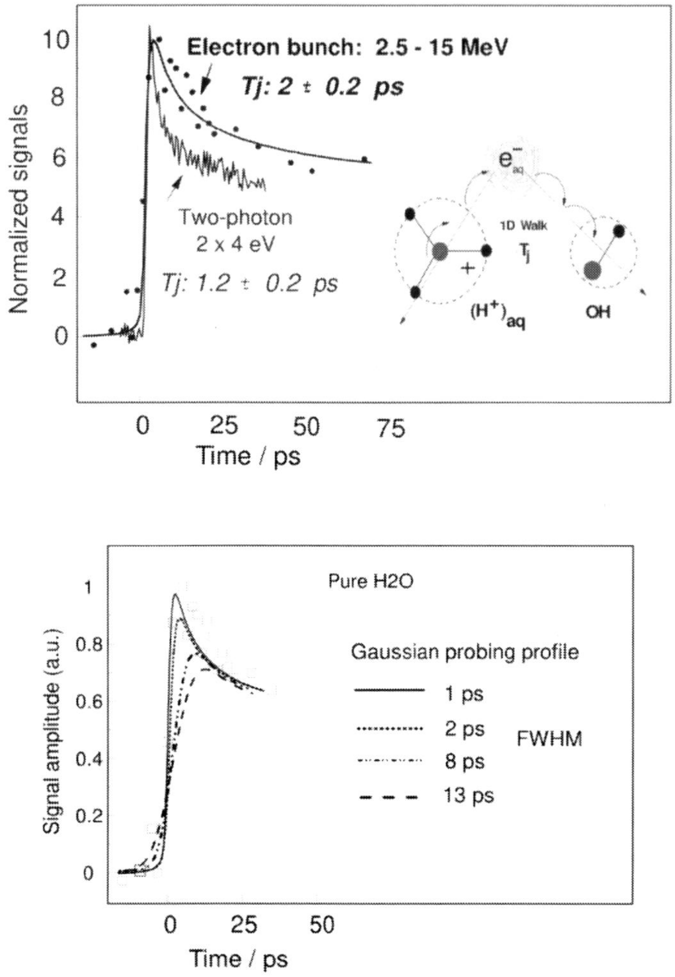

Figure 4. Upper part: time-dependence of fully hydrated electron population (e^-_{aq}) following the ionisation of liquid water by femtosecond electron bunches in the energy range 2.5-15 MeV or by a low-energy two-photon excitation process (8 eV). The 820 nm spectroscopic signature of hydrated electrons is characterized by dispersive 1D recombination processes (equation 3). In native ionization tracks the probability of early electron recombination with water products (γ =0.65) is higher than for low-energy two-photon excitation (γ =0.45). The incomplete signal decay is assigned to long-lived ground states of the hydrated electron.
Lower part: Influence of full width half magnitude (FWHM) of Gaussian probing profiles at 820 nm on the shape of early femtolysis signals following the interaction of femtosecond electron bunches (2.5-15 MeV) with a liquid water bath.

Femtolysis experiments emphasize that in native ionisation tracks, ultrafast geminate recombinations are dependent on the transient distribution of electron-hole pairs and the spatial configurations of electron-prototropic entities. A limit case corresponds to very short-range electron-proton couplings for which H_3O^+ undergoes one jump (1D motion by a finite process). In this way, considering a diffusion coefficient of H_3O^+ expressed as $D \sim \lambda d^2/6$ and an experimental jump frequency $\lambda = 1/Tj = 0.5 \times 10^{12}\ s^{-1}$, the initial proton jump distance would be around 2.8 Å. Femtolysis experiments also underline the significant contribution of highly reactive water bridged three-body complexes [$OH^\bullet ... e^- ... H_3O^+$] in early stages. As shown in the lower part of figure 4, the real-time discrimination of these non-independent pairs requires short FWHM Gaussian probing profiles, typically less than 2×10^{-12} s.

3. Modulation of early radical processes in native ionisation tracks

In nascent spurs, the escape probability of low energy secondary electrons is significantly dependent on ultra-fast recombination with two neoformed prototropic entities: OH and H_3O^+. These primary radical events influence also the early yield of low-energy localized electrons [16]. Laser-plasma accelerator based HERF may allow the direct observation of primary radiation events as a function of local environments induced by the presence of ionic entities or bio-molecules in the aqueous bath (figure 5). Indeed, when the high density of hydrogen bonding is significantly broken by a ionic strength of sodium chloride (H_2O/NaCl = 9), the contribution of early recombination processes is significantly reduced. This observation is assigned to specific field effects of anions and cations on the sub-structure of native ionization tracks; the breaking of the hydrogen network and a change of the density state fluctuation of the water bath would disturb the early spatial distribution of [$OH^\bullet \ldots e^- \ldots H_3O^+$] complexes and their ultrafast recombination [17,18]. HERF studies strongly suggest the pre-eminence of quantum effects during the sub-picosecond recombination of non-independent radical pairs within a fluctuating hydrogen bond network [16].

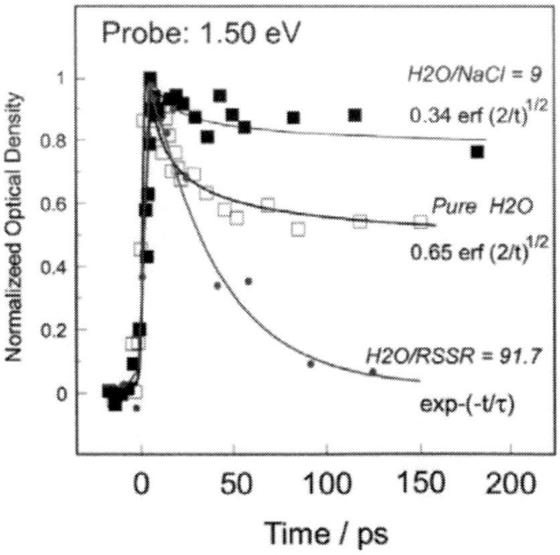

Figure 5. Influence of ionic and bio-molecular environments on the time dependence of 1s ground state hydrated electrons in nascent ionisation tracks. For these three conditions, the tracks are produced by femtosecond relativistic electron bunches of 2.5-15 MeV. The radiation chemistry processes are probed at 820 nm (1.5 eV)

With reference to femtolysis experiments performed with homogeneous aqueous samples, the dynamics of early events can also be modulated by molecules of biological interest such as the disulfide compound (NH_3-$(CH_2)_2$-SS-$(CH_2)_2$-NH_3) [19,20]. In the chemical core induced by femtosecond electron bunches, this disulfide compound favors a bimolecular reaction: an electron-biomolecule attachment from 1s-ground state hydrated electrons, characterized by a pseudo-first order dynamics (exp $(-t/\tau)$ with $\tau = 45$ ps). This diffusion-controlled process leads to the formation of sulfur-centered radical anions $RS{\therefore}SR^-$ characterized by a two-centre-three electron bond [21,22]. Indeed, small bio-molecules can be used as a sensor of specific environments determined by confined ionisation spaces. Future femtolysis developments would permit to extend our understanding of very short-time radical events in the prethermal regime of ionization tracks i.e. typically in the 10^{-14} -10^{-12} s window (figure 2).

4. HERF and prethermal radiation biomedicine at the nanometric scale

One of the most promising developments of future experimental HERF research will concern $^{i)}$ the real-time investigation of early radical events in the prethermal regime of native ionization tracks, considering short-lived quantum probes produced at the frontier of physical core and penumbra zone, $^{ii)}$ the quantification of ultrafast sub-nanometric bio-molecular damage (bond weakening and bond breaking) in the radial direction of relativistic particle beams, using the infrared spectroscopic properties of the very-short lived p-like configuration of excited electron [23]. This p-like excited state is the direct precursor of 1s ground state hydrated electrons [24-27]. It provides a ubiquitous sub-nanometric probe to explore short-time interactions with bio-molecules [22,28,29] and would offer the opportunity to characterize pertinent bio-effect parameters in native ionization tracks, as a function of the radiation quality (energy, time, fluence density and dose delivery rate profiles).

Compared to classical dose rate delivery generally used for conventional radiotherapies, ~ 1 Gy min^{-1}, the very high dose rate delivered with laser plasma accelerators, ~10^{13} Gy s^{-1}, may challenge our understanding of radiation effects on molecular targets, at the time scale of i) molecular motions, i.e. angstrom or sub-angstrom displacements, ii) radical and molecular vibrations. Regarding more integrated living systems such as normal and cancerous cells, the consequence of high dose rate profiles delivered by laser-plasma accelerated particle beams (electron, proton) requires the developments of bio-dosimetry concepts at the nanometric levels. Recent advances for spatio-temporal radiation biology are in progress [30-33]. They are a prerequisite for the biomedical applications of ultrashort radiation sources such as the real-time radio-sensitisation of tumour cells with vectorized nanoparticles or pulsed cancer chemoradiation therapies.

Acknowledgements
The author thanks several members of the *Radiation Biology* MELUSYN Network for fruitful discussions.

References

[1] Hendee W R, Ibbott G S and Hendee EG 2005 *Radiation therapy physics*, Wiley-Liss Ed. Von Sonntag C 2006 *Free-radical-Induced DNA Damage and its Repair* Heidelberg: Springer
[2] Sabin J R and Brandas E 2007 *Advances in quantum chemistry: theory of the interaction of radiation with biomolecules* Amsterdam, Elsevier
[3] Friedland W, Paretzke H G, Ballarini F, Ottolenghi A, Kreth G and Cremer C 2008 *Radiat. Environ. Biophys* **47** 49
[4] Shukla M and Leszczynski J 2008 *Radiation induced molecular phenomena in nucleic acids: a comprehensive theoretical and experimental analysis*, Springer
[5] Gauduel Y A (Editor) 2010 *Spatio-temporal radiation biology: Transdisciplinary advances for biomedical applications*. Mutation Research Reviews, **704** 214 page
[6] Faure J, Glinec Y, Pukhov A, Kiselev S, Gordienko S, Lefebvre E, Rousseau J P, Burgy F and Malka V 2004 *Nature* **431** 541
[7] Faure J, Rechatin C, Norlin A, Lifschitz A, Glinec, Y and Malka V. 2006 Nature **444**, 05393
[8] Oulianov D A, Crowell R A, Gosztola D J, Shkrob I A, Korovyanko O J and Rey-de-Castro R C 2007 *J. Appl. Phys.* **101** 053102-1-9
[9] Malka V, Faure J, Gauduel Y, Lefebvre E, Rousse A and Ta Phuoc K 2008 *Nature Phys.* **4** 447
[10] Tajima J, Habs D and Yan X 2009 *Rev. Acc. Sci. Tech.* **2** 201
[11] Gauduel Y A 2008 *Femtochemistry : lasers to investigate ultrafast reactions*. In *"Lasers in Chemistry"*, M. Lackner Ed, *VCH* **2**, 861-898
[12] Baxendale J H and Busi F 1982 *The Study of Fast Processes and Transient Species by Electron Pulse Radiolysis*, Dordrecht: Reidel Publishing Company
[13] Farhataziz, M A J and Rodgers 1987 *Radiation Chemistry; Principles and Applications*, Weinheim VCH
[14] Rechatin C, Faure J, Ben-Ismail A, Lim J, Fitour R, Specka A, Videau H, Tafzi A, Burgy F, and Malka V 2009 *Phys. Rev. Lett.* **102** 164801
[15] Esarey E, Schroeder C B and Leemans W P 2009 *Rev. Mod. Phys.* **81** 1229
[16] Gauduel Y A, Glinec Y, Rousseau J P, Burgy F and Malka V 2010 *Eur. Phys. J.* D **60** 121
[17] Cramer C J and Truhlar D G 1994 *Structure and reactivity in aqueous solution*, ACS Symposium, Washington, USA
[18] Gauduel Y, Gelabert H and Ashokumar M. 2005 *J. Mol. Liq.* **64** 57
[19] Bergamini A, Capozzi M, Ghibelli L, Dini L, Salanitro A, Milanese G, Wagner T, Beninati S, Delfina C D, Amici C and Rocchi G 1994 *J. Clin. Invest.* **93** 2251
[20] Oiry J, Pue J Y, Imbach J L, Fatome M, Sentenac-Roumanou H and Lion H 1986 *J. Med. Chem.* **29** 2217

[21] Alfassi Z.B. (Ed) 1998 *General aspects of the chemistry of radicals*, Wiley & Sons, Chichester
[22] Gauduel Y, Gelabert H and Guilloud F 2000 *J. Am. Chem. Soc.* **122** 5082
[23] Migus A, Gauduel Y, Martin J L and Antonetti A 1987 *Phys Rev Lett* **58** 1159
[24] Schnitker J and Rossky P J 1987 *J. Chem. Phys.* ***86*** 3471
[25] Wallqvist A, Martyna G and Berne B J 1988 *J. Phys. Chem* **92** 1721
[26] Schwartz B J and Rossky P J 1994 *J. Chem. Phys.* **101** 6902
[27] Gauduel Y, Hallou A and Charles B 2003 *J. Phys. Chem.* A **107** 2011
[28] Wang JCR, Nguyen J and Lu Q B 2009 *J. Am. Chem. Soc.* **131** 11320
[29] Sanche L 2009 *Nature* **461** 358
[30] Malka V, Faure J and Gauduel YA 2010 *Mut.. Res. Rev* **704** 142
[31] Yogo A *et al.* 2009 *Appl. Phys. Lett.* **94** 181502
[32] Rigaud O, Fortunel N O, Vaigot P, Cadio E, Martin M T, Lundh O, Faure J, Rechatin C, Malka V and Gauduel Y A 2010 *Cell Death Disease* **1** e73.
[33] Gauduel Y A 2011 *J. Phys. CS* **261** 012006

Journal of Physics: Conference Series **373** (2012) 012013 doi:10.1088/1742-6596/373/1/012013

Early models of DNA damage formation

Małgorzata A. Śmiałek[1,2]

[1]Department of Atomic Physics and Luminescence, Faculty of Applied Physics and Mathematics, Gdańsk University of Technology, Gabriela Narutowicza 11/12, 80-233 Gdańsk, Poland
[2]Department of Physical Sciences, The Open University, Milton Keynes MK7 6AA, United Kingdom

E-mail: smialek@mif.pg.gda.pl

Abstract. Quantification of DNA damage, induced by various types of incident radiation as well as chemical agents, has been the subject of many theoretical and experimental studies, supporting the development of modern cancer therapy. The primary observations showed that many factors can lead to damage of DNA molecules. It became clear that the development of experimental techniques for exploring this phenomenon is required. Another problem was simultaneously dealt with, anticipating on how the damage is distributed within the double helix of the DNA molecule and how the single strand break formation and accumulation can influence the lethal double strand break formation. In this work the most important probabilistic models for DNA strand breakage and damage propagation are summarized and compared.

1. Introduction

Although for many years DNA damage has been subject to various studies, in recent years a new approach has been undertaken in order to assess the influence of radiation and chemical agents action on DNA. Breakage of the DNA double helix has been investigated under various conditions when exposed to both ionizing and non-ionizing radiation as well as other reactive species, like hydroxyl radicals or cutting and nicking enzymes. In all these attempts the main goal was to be able to obtain highest levels of double-strand breaks (DSBs), lethal to living cells. It was also found that accumulation of single strand breaks (SSBs) can lead to DSBs appearance and further to cell death, when the single-strand lesions appear in both strands within a small distance, h. This distance is measured in basepairs for double stranded DNA molecule. It describes a span between SSBs created in opposite strands that will lead to hydrogen bond rupture between complementary bases lying between these breaks and further create a DSB in DNA molecule. Together with changes in DNA conformation, this distance became one of the most important parameters to be determined in samples exposed to damaging agents of various kinds.

Simultaneously, investigations on the probability of breaking a DNA strand by incident particles were conducted. In early years the damage distribution assessment was based on probabilistic models. The very first studies accounted for neither parameters like random lesion distribution within both strands nor the strength of phosphodiester bonds in the DNA backbone [1], and the experimental data fitted to the theoretical model that was developed gave a value of $h = 2$. Some later developments brought more light to the problem of how single-strand lesions are deployed between strands of DNA molecules and their influence on DSBs formation [2] as well as the difference in the enzyme and particle action on DNA molecule. Moreover, it was found that the conformation of the plasmid molecule has a great impact on the yields of damage obtained from irradiation [3].

The earliest studies employed sucrose gradient centrifugation for breaks analysis and focused only on transitions from relaxed, *R*, to linear, *L*, DNA form. When a new experimental method, agarose gel electrophoresis, was brought in for analysis of DNA damage induced by radiation [4], the possibility of investigating the decrease in supercoiled, *S*, DNA levels arose. The relations between various topological forms of DNA as well as levels of damage that could be detected with this new method were derived. Further fragmentation of linear molecules was also investigated and the loss of linear molecules due to multiple DSBs (MDSBs) formation was described [5]. This fragmentation was also considered with respect to the DSB in DNA being a result of a multiple SSBs accumulation or by simultaneous cleavage of both strands by a damaging agent.

In all cases the experimental results obtained for various types of damaging agents showed some discrepancies in the determination of the h value. In addition, with development of the theory, more accurate modeling of experimental data was possible, leading to a correct description of the damage formation mechanism.

Nowadays more sophisticated computational models are in use. Due to their complexity it is now possible to deal not only with simple break formation, but also to account for the influence of irradiation of the DNA environment, i.e. living cell, and thus secondary species attack [6–9]. Moreover, an accurate examination of the mode of action of various incident particles can be performed at the same time [10, 11].

The very first attempts to describe transformations in DNA conformations upon damage turned out to give a solid base for SSBs and DSBs determinations and are continuously used for simple modeling of damage in experimental works dealing with DNA and protein damage induced by chemical agents [12], ions [13], OH radicals [14, 15], photons [16] or neutrons [17]. In this paper the most important early developments in DNA damage assessment theory will be presented and summarized.

2. Modeling of DNA damage

Very early studies [1] introduced a simple model for the assessment of the h parameter from the molecular weight decrease of enzymatically treated DNA molecules. The main disadvantage of this approach was that it takes an average number of DSBs per DNA molecule only. Also, an assumption was made, stating that all phosphodiester bonds in a DNA molecule are equally breakable. Such simplifications could be valid when breaks are created by enzymes, such as DNA-ase, but they are not valid when particles are the damaging agent. Therefore, in order to satisfy more general experimental conditions, more detailed modeling was required.

2.1. The Freifelder and Trumbo model [2]

In the model developed by Freifelder and Trumbo [2] the authors analyzed the transitions between R and L DNA forms, initiated by X-ray radiation. Both the average numbers of DSBs and SSBs were taken into account in order to determine the h value. The reasoning for this work was a mismatch between new experimental data and earlier theory. The discrepancy was attributed to the effects caused by obtaining lesions in DNA molecules with irradiation by high energy particles rather than enzymes. By analyzing the probabilities of DSB formation depending on the position of the first and all the following SSBs in both strands, a general formula for the probability that no DSBs will occur in the molecule, $\mathscr{F}(N)$ was derived. By solving this equation, the authors showed that a fraction of molecules, which does not suffer a DSB is

$$\mathscr{F} = \left[\frac{\mathscr{L}-\mathscr{R}}{\mathscr{L}}\right]^b = \frac{\mathscr{L}-b(2h+1)^b}{\mathscr{L}}, \tag{1}$$

where $b=\frac{1}{2}p_s$ is the average number of SSBs per strand, p_s is the average number of SSBs per double-stranded molecule, $\mathscr{L}$ is the number of phosphodiester bonds per single strand and $\mathscr{R}$ is the number of forbidden positions for next SSB formation.

Using the Poisson distribution of SSBs in the molecule to evaluate their number N and thus accounting for varying numbers of such lesions in molecules that are present in the sample, a more precise equation describing $\mathscr{F}$ was obtained

$$\mathscr{F} = \sum_{N=0}^{\infty} [\mathscr{F}(N)e^{-p_s}(p_s^N/N!)], \tag{2}$$

where $\mathscr{F}(N)$ is a probability that N SSBs in a molecule are arranged so that no DSBs can be formed. The authors did not find this equation to be more precise when fitting their experimental data, in addition, the solution in computationally more difficult to find. Thus, the equation (1) was rearranged in order to obtain the value of h

$$h = \frac{\mathscr{L}}{2b}(1 - \mathscr{F}^*) - \frac{1}{2}, \tag{3}$$

where $\mathscr{F}^*$ is the b-th root of $\mathscr{F}$. If h needs to be expressed in terms of p_d, the average number of DSBs per molecule, the calculations lead to

$$h = \frac{p_d \mathscr{L}}{2b^2} - \frac{1}{2} \quad \text{or} \quad p_d = \frac{b^2}{\mathscr{L}}(2h+1). \tag{4}$$

Equation (4) is equivalent to the formula derived in [1], given for a polydisperse system.

To account for different probabilities of damaging phosphodiester bonds by irradiation the authors discussed the likelihood of the SSB occurrence next to a specific DNA base. Assuming that the DNA sample used in the experiment contained molecules large enough to be considered possessing a random sequence of nucleotides in both strands and that the breaks can only occur adjacent to a purine, a quarter of the phosphodiester bonds could not be broken. Therefore, the probability that a phosphodiester bond will be broken, $b/\mathscr{L}$, should be taken as $b/(3/4\mathscr{L})$ instead. As the same reasoning would apply to the number of forbidden regions in the molecule, $\mathscr{R}$, thus their ratio in equation (1) remains unchanged.

2.2. The Blok and Loman model [18]

In some later work the damage yield was analyzed in terms of the action of secondary particles, like electrons or radicals that are formed in living cells upon irradiation. First, a formula for D_{37}, a 37 % survival dose where, on average, there is one lethal lesion present per one DNA molecule, was derived to yield

$$D_{37} = \frac{100}{rQ}(N_0 + C), \tag{5}$$

where N_0 is the DNA concentration (number of molecules per ml), r is the number of eV per gram per rad. Q and C are given by $Q = \sum_i \varepsilon_i G_i$ and $C = \frac{S}{Q}\sum_i \alpha_i \varepsilon_i G_i$, where G_i is the radical yield (OH$^\bullet$, H$^\bullet$ or e^-_{aq}, respectively), and ε_i is the efficiency of inactivation of a reaction between the i-th radical and the DNA molecule. Because the solution always contains impurity molecules that remove part of the primary radicals, the α_i parameter, which describes a ratio of rate constants of the i-th radical for reaction with impurity and DNA molecules, respectively, was introduced. Nonetheless, no allowance for impurities present in the solution that originate from DNA preparation was made. Furthermore, the authors assumed that these impurities can only act as radical scavengers. The authors did not observe a linear relationship between D_{37} and DNA concentration in their experimental results. The discrepancies between the derived equation (5) and experimental data were explained to be due to a secondary radical formation in the DNA molecule upon attack of primary radicals. The molecule itself would not suffer damage but would interact further with other, undamaged molecules in the solution.

The average number of single strand breaks present in the double stranded molecule, p_s, being a sum of such lesions present before, p_s^0, and after irradiation, p, in a molecule containing $\mathscr{L}$ base pairs was shown to be

$$p_s = p + p_s^0 = 2\mathscr{L}n, \tag{6}$$

where n is the probability that a break has occurred between two arbitrarily chosen neighboring bases in a strand. The authors determined the average number of DSBs per molecule, p_d, with respect to the average number of SSBs that appear in the double stranded molecule

$$p_d = \frac{(p+p_s^0)^2}{4\mathscr{L}}(2h-1) = \frac{p^2+2p_s^0 p}{4\mathscr{L}}(2h-1)+const. \tag{7}$$

If no SSBs are present in the DNA sample before irradiation, the constant in equation (7) is small and can be neglected, thus the relationship is similar to (4) given in [2]. The obtained relationship contains both linear and quadratic terms for SSBs appearance in the sample upon irradiation. Also, in case of lack of initial breaks the formula simplifies to a quadratic-only relationship and h can be easily calculated from the slope of p_d versus p_s plot.

The value of h obtained by the authors was very close to the ones obtained previously, but a mismatch between calculated and measured values of p_s^0 gave a reasonable doubt on the accuracy and the rationale of theoretical analysis presented by the authors. Also, the experimental data were showing a linear increase in p_d with p_s at low doses of radiation, instead of a quadratic one. This was explained by introducing two mechanisms of DSBs formation. It was postulated that at low doses DSBs are formed by single-hit events involving radical attack and, thus, the initial presence of SSBs is not required, whereas at higher doses the accumulation of SSBs will cause an increase in the number of DSBs as a primary mechanism.

This formula was later used by van der Schans [3] giving four times larger values of h when applied to his data for γ-irradiated DNA samples. A temporal local denaturation following SSBs appearance was given as a possible explanation.

2.3. The van Touw model [4]

The model presented by van Touw and coworkers [4] was developed to handle the data authors obtained from a newly developed method for DNA breaks analysis - agarose gel electrophoresis. In their experiments they could analyze not only the transitions between R and L forms of DNA, but also the loss of the initially supercoiled DNA. Contrary to gradient centrifugation techniques [3, 19], for the first time it was possible to follow dose-dependent quantitative changes in all three DNA forms since they got separated completely on the gel. All the previously derived models either did not account for the contribution of various mechanisms to DSBs formation or treated the formation of SSBs by simply averaging their amount between both strands in the double stranded molecule. According to the authors, such an assumption was not correct because it does not allow for overlapping of SSBs in both strands within the h distance. The formulas derived here took into account a binomial distribution of the SSBs between two DNA strands and the probability that regions with SSBs in opposite strands within h can overlap each other.

The authors analyzed the probabilities with which the molecules in an irradiated sample are affected by SSBs and DSBs formation. Also, a multiple hit DSB originating from two independently formed SSBs was distinguished. Moreover, the model allows predicting levels of potential SSBs that are not visible on the gel, but still present in DNA molecules. For all these cases a binomial distribution of probability of a hit was used and a Poisson probability of lesions distribution was applied. The analysis led to the conclusion that a probability, w, that a SSB in one strand will find a SSB in the other strand within a distance h is $2a/\mathscr{L}$. Thus, a fraction of molecules, f, of contour length $\mathscr{L}$, that has, by single hits, n_d DSBs and n_s SSBs and additional n_{ds} DSBs as a result of n_s SSBs was derived to be

$$f(p_d,p_s,2h/\mathscr{L};n_d,n_s,n_{ds}) = P(p_d;n_d)\cdot\left\{P(p_s;n_s)\cdot\sum_{k_s=0}^{n_s}\left[B(n_s,1/2;k_s)\cdot B(m,w;n_{ds})\right]\right\}. \tag{8}$$

Noticing that f given by (8) is a multivariate distribution function, a relationship between p_{ds} and p_s (in

Journal of Physics: Conference Series **373** (2012) 012013 doi:10.1088/1742-6596/373/1/012013

agreement with [18]) was also obtained

$$p_{ds} = \left(\frac{p_s}{2}\right)^2 \cdot \frac{2h}{\mathscr{L}}. \tag{9}$$

The p_s and p_d parameters, representing the average numbers of SSBs and DSBs that originated from a single-hit event per molecule, can be related to the radiation dose, D, in the following way

$$p_s = p_s^0 + D/D_{37s}, \tag{10}$$

$$p_d = p_d^0 + D/D_{37d} = p_d^0 + \varepsilon p_s, \tag{11}$$

where p_d^0 and p_s^0 represent numbers of SSBs and DSBs in molecules at zero dose, respectively, and D_{37s} and D_{37d} are doses, at which there is, on average, one single-hit SSB and one single-hit DSB per DNA molecule, respectively. The ε parameter expresses the efficiency with which a single-hit DSB is created over a SSB.

From (8) the authors derived a set of formulas allowing to follow the changes in S, R and L forms of DNA. For fraction S that has no SSBs ($n_s = 0$) and DSBs ($n_d = 0$) and thus no multiple-hit DSBs ($n_{ds} = 0$)

$$S(n_d, n_s, n_{ds} = 0) = P(p_d;0) \cdot P(p_s;0). \tag{12}$$

In case of fraction R that has at least one SSB ($n_s \geq 1$), no DSBs ($n_d = 0$) and thus no multiple-hit DSBs ($n_{ds} = 0$), a following relation was derived

$$R(n_d, n_{ds} = 0, n_s \geq 1) = P(p_d;0) \cdot \left\{ \sum_{n_s=0}^{\infty} \left[P(p_s;n_s) \cdot \sum_{k_s=0}^{n_s} (B(n_s, 1/2; k_s) \cdot B(m, w; 0)) \right] - P(p_s;0) \right\}. \tag{13}$$

The last fraction, L, is the fraction, which has only one DSB created either by single- or double-hit ($n_d + n_{ds} = 1$)

$$L(n_d + n_{ds} = 1) = P(p_d;1) \cdot \sum_{n_s=0}^{\infty} \left\{ P(p_s;n_s) \cdot \sum_{k_s=0}^{n_s} [B(n_s, 1/2; k_s) \cdot B(m, w; 0)] \right\}$$
$$+ P(p_d;0) \cdot \sum_{n_s=0}^{\infty} \left\{ P(p_s;n_s) \cdot \sum_{k_s=0}^{n_s} [B(n_s, 1/2; k_s) \cdot B(m, w; 1)] \right\}. \tag{14}$$

The authors derived also a simplified set of equations describing quantitative changes in all three DNA fractions upon irradiation and lesions formation

$$S = \mathrm{e}^{-p_d} \cdot \mathrm{e}^{-p_s}, \tag{15}$$

$$R \simeq \mathrm{e}^{-p_d} \cdot \left[\mathrm{e}^{-p_{ds}} - \mathrm{e}^{-p_s}\right], \tag{16}$$

$$L \simeq (p_d + p_{ds}) \cdot \mathrm{e}^{-p_d} \cdot \mathrm{e}^{-p_{ds}}. \tag{17}$$

These equations are valid only when numbers of p_s and $2h/\mathscr{L}$ are low and p_d/p_s is high. Otherwise some inconsistency at higher doses, i.e. higher p_s numbers, can be seen between equations (12)–(14) and (15)–(17).

The approach taken by the authors allowed for the first time to determine numbers of SSBs, DSBs, loss of initial S form and the distance h, required to create a DSB from two independent SSBs in irradiated molecules. The distance h, calculated using this model, was shown to be approximately 30 bp. Such a large discrepancy with values found in the literature (see section 2.2) was ascribed to a difference in the DNA preparation method rather than a result of more accurate modeling. Although the authors provided a very detailed set of equations describing changes in topoisomeric forms of DNA (12)–(14), unfortunately the formulas did not include the evolution of the fragmented form, F, which certainly appears at higher doses of radiation.

Nonetheless, it was also possible to follow the time evolution of three main DNA topoisomeric forms. The simplicity of the new experimental method – agarose gel electrophoresis – made it being widely used by scientists in radiation studies.

This model was later modified by Hempel and Mildenberger [20], who noticed that the number of DSBs, p_d given by (11) should include a factor for p_{ds} as both types of damage will create an L form of DNA

$$\widetilde{p}_d = p_d + p_{ds}. \tag{18}$$

The h value obtained from fitting to the experimental data was much higher than reported by others, reaching 60 bp. It was doubted that such long distance between SSBs would cause enough instability in the molecule in order to create a DSB. The authors had also noticed that from the simplified set of equations (15)–(17) one can derive another formula describing changes in the levels of DSBs, namely

$$\widetilde{p}_d = \frac{L}{S+R}. \tag{19}$$

The advantage of this modification was that only the relative fluorescence measurements of the gel had to be performed in order to assess levels of DNA damage.

2.4. *The Cowan model [5]*

Simultaneously, another model, based on earlier work [2, 21], was derived. The damage to the DNA sample was introduced by nicking and cutting enzymes that can mimic breaks appearance in DNA upon irradiation, allowing to fully control damage induction. Apart from managing the transitions between S, R and L DNA forms, the authors also considered various cases, including further fragmentation of linear DNA due to MDSBs formation and thus appearance of the F form. In addition, a situation, where not 100 % of the starting material is in a supercoiled form was discussed.

First, the case with only the single stranded nicking being the damaging agent was considered. Changes in levels of the supercoiled DNA form were assumed to follow an exponential loss of the starting material with time, t

$$S(\mu) = \mathrm{e}^{-\mu}. \tag{20}$$

with $\mu = \lambda \cdot t$, where λ is a proportionality constant. The μ parameter is the number of nicks created in a molecule up to time t and can be considered as a 'nicking dose'. Derivation of other formulas, describing changes in levels of R and L forms, required more advanced mathematics and led the authors to the following relations

$$R(\mu) = 2\mathrm{e}^{-\mu/2} - 2\mathrm{e}^{-\mu} + \mu X, \tag{21}$$

$$L(\mu) > q^{-1}\left(\mathrm{e}^{\mu q/2} - 1\right)\left(\mu X - Y + \mathrm{e}^{-\mu/2} - \mathrm{e}^{-\mu}\right) \qquad \text{and} \tag{22}$$

$$L(\mu) < \mu\,(2 - q\mu)^{-1}\left(\mu X - Y + \mathrm{e}^{-\mu/2} - \mathrm{e}^{-\mu}\right), \tag{23}$$

where

$$X = \sum_{k=1}^{\infty} \mathrm{e}^{-\mu(1+kq)/2}[\mu(1-kq)_{+}/2]^{2k-1}/(2k!) \qquad \text{and} \tag{24}$$

$$Y = \sum_{k=1}^{\infty} \mathrm{e}^{-\mu(1+kq)/2}[\mu(1-kq)_{+}/2]^{2k-1}[2k + \mu(1-kq)/2]/(2k!). \tag{25}$$

Journal of Physics: Conference Series **373** (2012) 012013 doi:10.1088/1742-6596/373/1/012013

The q parameter denotes the size of the taboo zone – the area, where no further breaks can form. The MDSBs were simply assessed to be the rest of the material missing from the quantification of the three main forms with respect to the initial conditions

$$F(\mu) = 1 - S(\mu) - R(\mu) - L(\mu). \tag{26}$$

When an additional factor, causing a DSB formation simultaneously to SSBs appearance with efficiency ϕ (cutting dose), proportional to t via constant ρ, was introduced, the equations (20)–(23) took the following form

$$S(\mu,\phi) = \mathrm{e}^{-\phi} S(\mu), \tag{27}$$

$$R(\mu,\phi) = \mathrm{e}^{-\phi} R(\mu), \tag{28}$$

$$L(\mu,\phi) = \mathrm{e}^{-\phi}[L(\mu) + \phi\{S(\mu) + R(\mu)\}]. \tag{29}$$

In the situation, where the cleaving mechanism dominates over the single strand cut, it was possible to simplify the theory to

$$S(\mu,\phi) = \mathrm{e}^{-(\mu+\phi)} \tag{30}$$

$$R(\mu,\phi) = \left(1 - \mathrm{e}^{-\mu}\right)\mathrm{e}^{-\phi} \tag{31}$$

$$L(\mu,\phi) = \phi \mathrm{e}^{-\phi} \tag{32}$$

$$F(\mu,\phi) = 1 - \mathrm{e}^{-\phi}(1+\phi). \tag{33}$$

By dividing equations (30)–(32) by $1 - F(\mu/\phi)$, the authors obtained a set of formulas that correspond to the bands on gel and thus values for S, R and L can be obtained directly. In such case also μ and ϕ can be determined as

$$\phi = L'/(1-L'), \tag{34}$$

$$\mu = -\ln[S'(1+\phi)], \tag{35}$$

where S' and L' are the relative proportions of S and L forms. Accounting for the fact that not 100 % of the starting material being in supercoiled form increased to the values of μ and ϕ by starting nicking and cutting doses μ_0 and ϕ_0

$$\mu = \mu_0 + \lambda t, \tag{36}$$

$$\phi = \phi_0 + \rho t. \tag{37}$$

The last situation that was analyzed was the case when both mechanisms were taking place together with topoisomerase I action that unwinds supercoiled molecules leaving them in an R form with no breaks. Therefore, it was necessary to introduce another parameter, θ, describing the rate of topoisomerase I interaction with DNA. This parameter is analogous to μ and ϕ, except that the interaction has an effect only if it is the primary one of all three. Equations (26) and (29) remain unchanged, whereas (27) and (28) take the following form

$$S(\mu,\phi,\theta) = \mathrm{e}^{-\theta} S(\mu,\phi), \tag{38}$$

$$R(\mu,\phi,\theta) = R(\mu,\phi) + \left(1 - \mathrm{e}^{-\theta}\right) S(\mu,\phi). \tag{39}$$

As previously, a simplified theory for the case of DNA cleavage being the predominant mechanism was derived. The obtained equations were the same, except for μ being replaced by $\mu + \theta$.

From plots of molecular proportions of different DNA forms, obtained for varied values of all three parameters, the authors concluded that if only the nicking mechanism of DNA damage is involved, the DSBs start appearing when almost all of the S form of DNA is lost. When, in addition, a cleaving mechanism is involved, even a small increase in the cleaving rate ϕ would cause much faster degradation of the starting material; the simplified theory can only be used when DNA cleaving is truly the predominant mechanism. Their final finding was that if in the initial sample there are some relaxed but not nicked molecules or the molecules, initially supercoiled, undergo unwinding before nicking and cleaving, the loss of the supercoiled material is much faster, yet the topo-enzyme plays no role once all supercoiled molecules are lost. In comparison to the earlier model [2], the authors used their exact formula (21) and concluded that the oversimplified approach used previously underestimates the proportion of relaxed molecules being lost due to DSBs formation from multiple SSBs.

3. Summary

In this paper a few initial statistical models aiming at describing the distribution of lesions in DNA molecules, caused by damaging agents like enzymes, ionizing radiation or secondary species attack, are presented. With the development of experimental techniques, allowing for the separation of main topological forms of DNA, there was an urge to improve models in order to account for new phenomena that could then be observed. The earliest attempts were developed further to account for more topological DNA forms as well as for various conditions under which damage can be induced. The value of h parameter varies between models and experiments, still raising the issue of the length of the h that is necessary to create a DSB from two SSBs.

The distinction between the mode of action of an enzyme and radiation on a DNA molecule permitted to develop independent methods for description of topological changes in DNA. The accurate description of dynamical changes in DNA conformations upon the action of the damaging agent allows an assessment of the efficiency of the agent in damaging the DNA molecule that is an important issue in cancer therapy. The most advanced models of those presented here have laid the foundation for more advanced modeling developed recently. Nonetheless, they are still in use, aiding the analysis of experimental data and dealing with outcomes of the radiation-induced damage by electrons, photons or ions.

Acknowledgments

I would like to acknowledge the visiting fellow position with the Open University and DS grant earmarked for training of young scientists from GUT.

References

[1] Thomas C A and Doty P 1956 *J. Am. Chem. Soc.* **78** 1854–60
[2] Freifelder D and Trumbo B 1969 *Biopolymers* **7** 681–93
[3] van der Schans G P 1978 *Int. J. Radiat. Biol.* **33** 105–20
[4] van Touw J H, Verberne J B, Retel J and Loman H 1985 *Int. J. Radiat. Biol.* **48** 567–78
[5] Cowan R, Collis C M and Grigg G W 1987 *J. Theor. Biol.* **127** 229–45
[6] Holley W and Chatterjee A 1996 *Radiat. Res.* **145** 188–99
[7] Deppman A, Echeimberg J, Gouveia A, Arruda-Neto J, Milian F, Added N, Camargo M, Guzman F, Helene O, Likhachev V, Rodriguez O, Schenberg A, Vanin V and Vicente E 2004 *Braz. J. Phys.* **34** 958–61
[8] Surdutovich E, Obolensky O I, Scifoni E, Pshenichnov I, Mishustin I, Solov'yov A V and Greiner W 2009 *Eur. Phys. J.* D **51** 63–71
[9] Scifoni E, Surdutovich E and Solov'yov A V 2010 *Phys. Rev.* E **81** 1021903:1–7
[10] Surdutovich E, Scifoni E and Solov'yov A V 2010 *Mutat. Res.-Rev. Mutat. Res.* **704** 206–12
[11] Baccarelli I, Bald I, Gianturco F A, Illenberger E and Kopyra J 2011 *Phys. Rep.-Rev. Sec. Phys. Lett.* **508** 1–44
[12] Zhang Q, Xiang Y, Liang D, Peng Y and Guo H 2012 *Bioorg. Med. Chem. Lett.* **22** 1814–7
[13] Dang H M, van Goethem M J, van der Graaf E R, Brandenburg S, Hoekstra R and Schlathölter T 2011 *Eur. Phys. J.* D **63** 359–67
[14] Smialek M A, Moore S A, Mason N J and Shuker D E G 2009 *Radiat. Res.* **172** 529–36
[15] Spotheim-Maurizot M and Davidkova M 2011 *Mutat. Res.-Fundam. Mol. Mech. Mutagen.* **711** 41–8
[16] Filippova E M, Monteleone D C, Trunk J G, Sutherland B M, Quake S R and Sutherland J C 2003 *Biophys. J.* **84** 1281–90
[17] Spotheim-Maurizot M, Charlier M and Sabattier R 1990 *Int. J. Radiat. Biol.* **57** 301–13
[18] Blok J and Loman H 1973 *Curr. Top. Radiat. Res. Q.* **9** 165–245
[19] Freifelder D 1965 *Proc. Nat. Acad. Sci.* **54** 128–34
[20] Hempel K and Mildenberger E 1987 *Int. J. Radiat. Biol.* **52** 125–38
[21] Povirk L F, Wubker W, Kohnlein W and Hutchinson F 1977 *Nucleic Acids Res.* **4** 3573–80

Journal of Physics: Conference Series **373** (2012) 012014 doi:10.1088/1742-6596/373/1/012014

Damage of DNA backbone by nanoscale shock waves

Alexander V. Yakubovich[1,2] **, Eugene Surdutovich**[3] **and Andrey V. Solov'yov**[1,2]

[1]Frankfurt Institute for Advanced Studies, Ruth-Moufang Str. 1, Frankfurt am Main D-60438, Germany

[2]On leave from A.F. Ioffe Physical Technical Institute, Polytechnicheskaya 26, Saint-Petersburg, 194021, Russia

[3]Oakland University, Department of Physics, Rochester, MI 48309 USA

E-mail: `yakubovich@fias.uni-frankfurt.de`

Abstract. By means of full-atom molecular dynamics simulations we investigate the process of DNA backbone damage by nanoscale shock waves in a water environment. These shock waves are created by ions penetrating the medium with high linear energy transfer. The high rate of the ions' energy transfer to the surrounding molecules leads to the rapid increase of the temperature in the vicinity of the ion trajectory, which causes the formation of shock waves propagating through the medium. We have investigated the ions' linear energy transfer of 900, 2000 and 5000 eV/nm. In the case of a linear energy transfer of 5000 eV/nm the deposition exerts unsustainable stress onthe DNA molecule, which leads to the breakage of DNA backbone covalent bonds by thermomechanical effects.

1. Introduction

The damaging effects of energetic radiation have been known for many years. It has been commonly accepted that high-energy tracks formed by photons, electrons and atomic ions ionise cell components along the track, thereby leading to various dissociation channels and to the formation of damaging radicals. This knowledge has triggered intensive research on radiation protection and the development of biomedical applications for different types of ionising radiation, generally referred to as radiotherapy, especially used for treatment of the tumorous diseases [1].

An event of DNA damage due to an incident ion is rather complicated. For example, secondary electrons produced by the projectile may directly interact with a DNA molecule and yield a lesion. Alternatively, such a lesion can be produced by the vacancy hole left by the ejected electron [2–5].

On the theoretical side, much work has investigated different mechanisms of DNA damage. From these studies, the direct effects are the most important pathways in connection to damage induced by ions [2, 3, 6, 7]. All these mechanisms are related to the interactions of low-energy electrons or holes with DNA molecules. Alternatively a purely physical mechanism of damage has been discussed in reference [8]. That work has been devoted to the heat transfer in the vicinity of ion tracks. The solutions to the coupled thermal conductivity equations for electronic and atomic components predicted short-time temperature spikes. It was argued that these spikes may contribute to other mechanisms of damage and affect the chemical environment of the DNA. Following that paper, the molecular dynamics simulations of a biomolecule in the predicted temperature regimes were carried out [9]. The thermal spikes form on a time scale of $10^{-15} - 10^{-13}$ s after the ion's traverse [8]. They result from the energy transfer from the secondary electrons to the vibrational degrees of freedom of molecules of the medium, which dominates the other channels of energy transfer [8, 10]. Therefore, the energy per unit length of ion's trajectory of the order of LET is deemed to be converted to the kinetic energy of the expanding medium. This expansion was considered in reference [11] and described as a cylindrical shock wave following a strong explosion at 10^{-13}s after the ion's traverse. The first results of simulations of the effects of a heat spike in the vicinity of a nucleosome were reported in reference [12]. Those simulations were carried out taking into account all atoms in the system. It was shown that a high rate of energy transfer from an ion to the molecules of surrounding water environment leads to the formation of the nanoscale shock wave propagating through the medium, the same as predicted in reference [11]. The formed shock waves hitting the DNA can substantially deform its secondary structure. In reference [12], the linear energy transfer (LET) of the ion was taken to be 900 eV/nm, which corresponds to the carbon ion in the vicinity of its Bragg peak. Using the simple Boltzmann formalism, the probability of DNA backbone breakage was calculated, and the number of such strand breaks per one cell nucleus was estimated.

This work continues the study of the possibility of producing DNA lesions as a result of a thermomechanical effect caused by ions stopping in the tissue. By means of molecular dynamics simulations we investigate the effects of higher LET. The LET is higher due to a higher charge of the projectiles. In general, LET is proportional to the square of this charge. Initially fully charged ions tend to pick up electrons from the medium, and therefore, the initial charge becomes smaller as the speed of the projectiles decreases. The effective charge defining LET may be larger for heavier ions. Alternatively, the charge might be higher than the effective charge stemming from fluctuations due to collisions with molecules of the medium. Indeed, when ions are propagated through the medium they experience multiple collisions with molecules of the environment. These collisions lead to the continuous process of ions' ionization and recharging by capture of the electrons from the surrounding molecules. We discuss this process in more detail in the following section.

In section 2, we give a description of the simulations and the theoretical formalism for calculating the effective charge of the ion. In section 3, we discuss the results before concluding in section 4.

2. Theoretical methods

The effective charge of an ion moving through a water medium can be evaluated by using the formalism developed by Barkas [13] as follows:

$$z_{eff} = z\left[1 - \exp\left(-125\beta z^{-2/3}\right)\right], \tag{1}$$

where z_{eff} is the effective charge of an ion, z is the charge of the nucleus, $\beta = V/c$, where V and c are the velocity of an ion and speed of light, respectively.

The Bragg peak region of the ion's trajectory corresponds to a kinetic energy of approximately 0.3 MeV per nucleon [14]. The velocity of the ion with this kinetic energy is $\sim 2.52 \cdot 10^{-2}c$. Substituting the ion's charge and velocity to equation (1) one obtains for the case of carbon ions $z_{eff} \approx 3.70$.

The LET of the ion is proportional to its squared effective charge [14]. In this work we have investigated the ions' LETs of 900 eV/nm, 2000 eV/nm and 5000 eV/nm, which correspond to carbon, magnesium and arsenic. The properties of the ions are summarized in table (1).

Note that equation (1) gives the mean effective charge of the propagating ion. However, the ion experiences multiple collisions with molecules of water. These collisions lead to continuous processes of ion recharging by capture of electrons from the surrounding molecules

Table 1. Values of investigated LET and corresponding elements. z_{eff} is the effective charge of an ion in its Bragg peak as follows from equation (1), z is the total charge of the ion nucleus.

LET (eV/nm)	z_{eff}	z	Element
900	3.70	6	C
1080	4.04	7	N
1257	4.36	8	O
2000	5.50	12	Mg
4072	7.86	26	Fe
5000	8.72	33	As
10527	12.64	82	Pb
11503	13.21	92	U

and ionizations. Therefore, the effective charge of ions fluctuates, and on some parts of the trajectory ions move with higher charge and, consequently, LET. For instance, if the charge of a carbon ion in the vicinity of its Bragg peak is +6 for certain parts of its trajectory, the LET there increases up to 2400 eV/nm. A more accurate analysis of these charge fluctuations is postponed for a future work.

Below we present the results of molecular dynamics (MD) simulations of the temperature spike in the vicinity of a nucleosome, i.e., a DNA fragment wrapped around a histone-protein complex. This unit is of particular interest in connection with a so-called complex DNA damage, which is deemed to be lethal to the cell [15]. The heat spike in the liquid water medium is caused by the propagation of a heavy ion [8]. We consider this effect at that part of the ion trajectory where the LET is the highest, i.e. at the so-called Bragg peak. In the vicinity of the Bragg peak the LET for carbon ions can be up to 900 eV/nm [8].

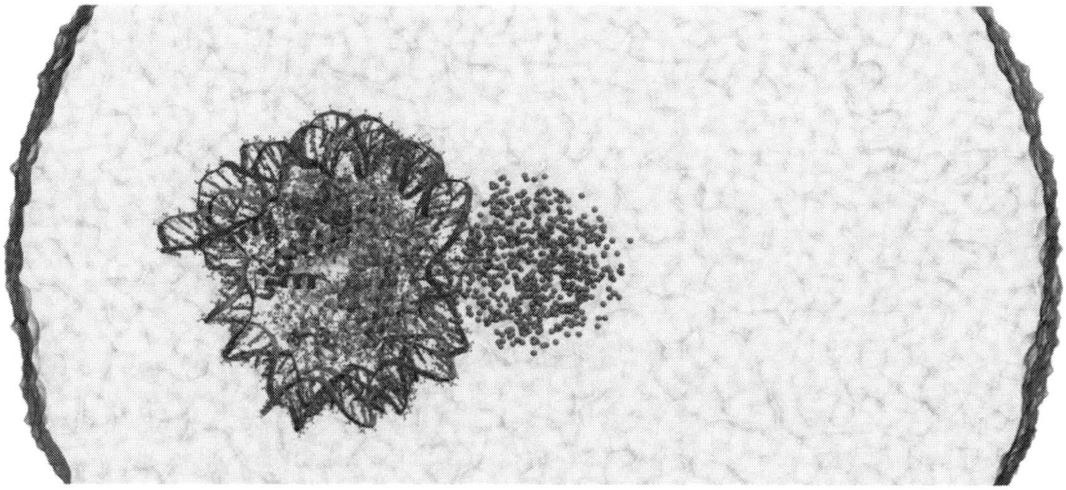

Figure 1. The strucutre of nucleosome solvated in a water cylinder 100 fs after heavy ion propagation. Water molecules located within 1 nm from ions track are shown by red spheres. All other water molecules in the system are shown by transparent blue surface.

The crystallographic structure of a nucleosome was taken from the Protein Data Bank (PDB) using the PDB ID 3LEL [16]. The structure of the nucleosome is defined by approximately one-half of the sequence of human α-satellite constructs NCP147 or NCP146b. These constructs yield the best diffracting crystals to date [17]. Obviously, different nucleosomes have a different sequence of constituting DNA fragments. In the present work we investigate the thermomechanical damage for one particular nucleosome structure assuming that the effects of thermomechanical damage are essentially very similar for other DNA fragments. The original structure in the PDB consists of two nucleosome fragments. Only the fragment with smaller values of the coordinates the along z-axis was chosen for MD simulations. The nucleosome was solvated in the water cylinder with the radius of 175 Å and 70 Å in length. Sodium chloride ions with the concentration of 0.15 mol/L were added to the solution and the total charge of the system was equal to zero. A nucleosome has a roughly disk-like shape. The axis of this "disk" was chosen to be parallel to that of the cylinder. The distance from the cylinder axis to the closest atom of the nucleosome was equal to 15 Å. Periodic boundary conditions were applied along the z-axis of the water cylinder and repulsive boundary condition were applied on the cylinders xy plane. After the optimization of the system using a conjugate gradient algorithm the system was equilibrated for 33 ps at 310 K using a Langevin thermostat with a damping constant equal to 5 ps^{-1}. All the MD simulations presented in this work were conducted using the NAMD software package [18] and the CHARMM22 [19] forcefield. Water molecules were simulated using the TIP3 model [19]. The cut-offs for Van-der-Waals and Coulomb interactions were chosen to be equal to 6 and 11.5 Å, respectively. The switching and pairlist distance for Coulomb interactions were chosen to be 10 and 12.5 Å, respectively. The long-range Coulomb interactions were treated using the Particle-Mesh Ewald algorithm with a box size of 380, 380, and 70 Å along x-, y-, and z-axes, correspondingly. The number of grids in the particle mesh Ewald algorithm was equal to 384, 384, and 96 along x, y, and z axes, correspondingly. It takes about 15 ps for the shock wave to reach the cylinder boundary for LET= 5000 eV/nm. Therefore, during the first 10 ps after the ion's passage the effects of scattering of the shock wave from the cylinder's boundaries can be neglected. In a real cell the volume of the nucleus is densely filled with DNA and protein molecules, and an event of observing only a single nucleosome within the radius of 175 Å form the ion's track is very unlikely. However, in the present study we investigate the damage of a nucleosome located only 15 Å from the ion's path. We assume that the presence of other than water molecules in the cell nucleus will have a minor influence on the process of thermomechanical damage.

In this work, we assumed that the deposition of the ion energy into the kinetic energy of water molecules occurs instantaneously. This assumption is applicable in the present work since the characteristic thermalisation time of secondary electrons is smaller than the time needed for the formation of the shock wave being of the order of 50 fs, see reference [20]. We also assumed that most of the ion energy is deposited evenly between water molecules, located not farther than 1 nm from ion trajectory, see [12] for justification of this assumption.

For the simplicity of notations, we will refer to a cylinder (coaxial with the ion trajectory) of radius 1 nm as *small* cylinder. Then, in order to simulate heating of the medium by the ion, we have increased the velocities of the water molecules in the vicinity of the ion track (see figure 1). The more detailed energy distribution within this hot cylinder would introduce minor corrections to the structure of the shock wave (whose strength was justified in reference [11]) and the energy transferred to the covalent bonds, studied in this paper. Therefore, the heating was modeled by the multiplication of the absolute values of velocities of water molecules in the small cylinder by a certain factor α, which was calculated from the following equation:

$$\sum_{i=1}^{N} \frac{(\alpha v_i)^2}{2m_i} = \frac{3}{2}NkT + l\frac{|dE|}{dx} , \tag{2}$$

where N is the total number of atoms in the small cylinder, v_i, m_i, k, and T are the velocity and mass of i-th atom, the Boltzmann constant, and the absolute temperature, respectively,

l and $\frac{|dE|}{dx}$ are the length of the cylinder and LET of the ion. In this paper $l = 7$ nm and $\frac{|dE|}{dx} = 900$, 2000 and 5000 eV/nm, which corresponds to the maximum value of the LET for carbon, magnesium and arsenic ions moving in water in the vicinity of the Bragg peak [9], correspondingly.

The molecular dynamics simulations of the heat spike were started using atoms' coordinates and velocities, recorded after equilibration. The velocities of atoms in the small cylinder were increased by a factor α derived in equation (2). The molecular dynamics simulations were carried out in the NVE microcanonical ensemble using periodic boundary conditions along the z-axis of the cylinder. The MD simulations were performed for the first 10 ps after the ion passage using an integration time step of 0.1 fs. We have performed five, three and three independent runs of MD for LET 900, 2000 and 5000 eV/nm, correspondingly.

3. Results and discussion

Let us analyse whether the thermomechanical stress in the DNA fragment caused by the propagation of a shock wave created by an ion can be sufficient to induce the thermomechanical breakage of the covalent bonds in the DNA backbone. The energy of the molecule covalent bonds in the molecular mechanics potential utilized in the present work is described using the following empirical potential, see [19]:

$$\begin{aligned} U &= \sum_{i=1}^{N_b} k_i^b (r_i - r_i^0)^2 + \sum_{i=1}^{N_a} k_i^a (\theta_i - \theta_i^0)^2 + \\ &+ \sum_{i=1}^{N_d} k_i^d \left[1 + \cos(n_i \phi_i + \delta_i)\right] + \sum_{i=1}^{N_{id}} k_i^{id} (S_i - S_i^0)^2 \end{aligned} \tag{3}$$

Here the four terms describe the potential energy with respect to variations of distances, angles, dihedral angles and improper dihedral angles between two, three and four neighbouring atoms respectively. The summation in the first term goes over all topologically defined bonds in the system, in the second term over all topologically defined angles, and in the third one - over all topologically defined dihedral angles and in the fourth - over all topologically defined improper dihedral angles.

It is possible to reconstruct the energy of stretching stored in each covalent bond in the system knowing the coordinates of all the atoms as a function of simulation time. This can be achieved using the equation:

$$U_i(t) = k_i^b \left(|\mathbf{r_1}(t) - \mathbf{r_2}(t)| - r_i^0 \right)^2, \tag{4}$$

where $\mathbf{r_1}(t)$ and $\mathbf{r_2}(t)$ are the coordinates of two atoms forming the $i-th$ covalent bond, k_i^b and r_i^0 are the stiffness and equilibrium distance of the $i-th$ covalent bond, as defined in equation (4), respectively. Figure 2 shows the energies of stretching of the DNA backbone covalent bonds in the residue closest to the carbon ion trajectory. The results of all independent runs of MD simulations are shown in corresponding graphs of figure 2 simultaneously. The energies are derived from equation (4) and the coordinates obtained from MD simulation. Figure 2 shows that for short periods of time the energy of the covalent bonds stretching in the system can be up to 4.4 eV for LET 5000 eV/nm. The energies of covalent bonds of DNA backbone are $\sim 3.5\ldots3.7$ eV, as follows from [21], where the energies of DNA covalent bonds were calculated using density functional theory. One can conclude, that the stress deposited by thermomechanical effects to the DNA backbone is *sufficient* to break the covalent bonds of the backbone for the case of LET 5000 eV/nm. Note that there is no preferential covalent bond of the DNA backbone to which the energy of the shock wave is deposited. Therefore, it is possible to conclude that the

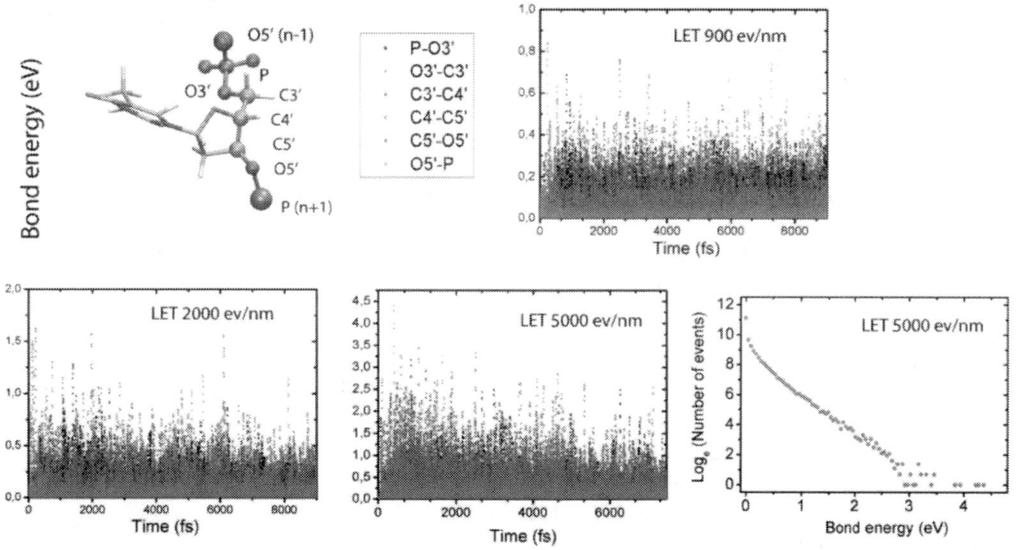

Figure 2. Energy of stretching of the covalent bonds in the DNA residue closest to the trajectory of carbon ion in the vicinity of the Bragg peak. The notation of different atoms in the residue is shown on the top-left of the graph. The energy of stretching of the bonds is plotted as a function of time calculated form the moment of ion passage. Top-right, low-left and low-center graphs correspond to ions LET 900, 2000 and 5000 eV/nm, respectively. The probability for observing a certain amount of energy in the covalent bonds in the DNA backbone is shown in the low-left graph.

stresses on the DNA backbone are homogeneously distributed among the bonds being at the same distance from the ion track. The molecular mechanics potential used in the present work does not allow for the breakage of covalent bonds in the system. However, the mean potential energy stored in covalent bonds of DNA backbone during first several pisoseconds after ion's passage is ~ 0.19 eV for the case of LET 5000 eV/nm. Since the energy of the covalent bond is several eV, only a very tiny fraction of the bonds in the molecules can be broken. Therefore, we assume that the effects of bond breakage will have a minor influence on the energy flow in the system, and will address this problem in grater detail in further work.

With the increase of time after the carbon ion passage the fluctuations of the energies of the covalent bonds stretching decrease since the pressure and the temperature in the wake of the wave front are substantially smaller than that at the front of the shock wave. In other words, the most thermomechanical stress to the biomolecule is caused at the moment of passage of the shock wave front through the molecule or directly after it.

In our work we have performed only 5, 3 and 3 simulations for LET 900, 2000 and 5000 eV/nm, correspondingly. For LET 900 and 2000 eV/nm the maximum observed bond energies are ~ 0.85 and ~ 1.6 eV, correspondingly. These values are lower than the DNA covalent bonds energies. However, within single cell nucleus there is a large number of events of ion's passage close to the DNA molecules. In reference [12], the number of events of ion's close passage was estimated as being of the order of a few hundreds. It is almost impracticable to perform several hundreds of simulations of heat spikes in order to directly observe the event of bond breakage. However, one can expect that with certain probability one can observe the events of DNA backbone damage by the thermomechanical effects for the cases of LET lower than 5000 eV/nm [12].

In summary, this paper demonstrates the importance of the thermomechanical mechanism of damage of biomolecules caused by the ion propagation through the biological medium. The reported mechanism of thermomechanical damage should be investigated in greater detail.

4. Conclusions

In this paper we have performed the MD simulation of a shock wave propagating through a nucleosome. The shock wave in the Bragg peak is caused by heating of the water medium during the ion propagation.

We have investigated the cases of LET 900, 2000 and 5000 eV/nm, which correspond to the effective charge of an ion equal to 2.55, 3.8 and 6, respectively.

The results of MD simulations demonstrate that the three-dimensional structure of the nucloeosome experiences noticeable deformations. The propagating shock wave distorts the stacking interactions and the hydrogen bonds stabilizing the nucleotides of the DNA and leads to the partial destruction of the DNA secondary structure.

For the case of LET 5000 eV/nm we have observed the events corresponding to deposition of 4.4 eV to certain bonds of the DNA backbone. 4.4 eV is higher than the energy of DNA covalent bonds. Therefore, it is possible to conclude that the DNA molecule cannot sustain such stresses and should be broken. In other words, we observe the event of DNA backbone damage by thermomechanical mechanism.

5. Acknowledgements

A.Y. thanks the Stiftung Polytechnische Gesellschaft Frankfurt am Main for its financial support and appreciates the use of resources at the Frankfurt Center for Scientific Computing. The authors thank the COST Action MP1002 (Nano-IBCT) for the support of the work.

References

[1] Baccarelli I, Gianturcol F, Scifoni E, Solov'yov A and Surdutovich E 2010 *Eur. Phys. J.* D **60** 1–10
[2] Becker D and Sevilla M 1993 *Advances in Radiation Biology* vol **17** ed Lett J (Acad. Press) pp 121–180
[3] Becker D, Adhikary A and Sevilla M 2010 *Charged Particle and Photon Interactions with Matter Recent Advances, Applications, and Interfaces* (Boca Raton: CRC Press, Taylor & Francis)
[4] Sanche L 2005 *Eur. Phys. J.* D **35** 367–390
[5] von Sonntag C 1987 *The chemical basis of radiation biology* (London: Taylor & Francis)
[6] Adhikary A, Kumar A and Sevilla M 2006 *Radiat. Res.* **165** 479–484
[7] Baccarelli I, Gianturco F, Grandi A, Lucchese R and Sanna N 2007 *Adv. Quantum Chem.* **52** 189
[8] Toulemonde M, Surdutovich E and Solov'yov A 2009 *Phys. Rev.* E **80** 031913
[9] Surdutovich E, Yakubovich A and Solov'yov A 2010 *Eur. Phys. J.* D **60** 101–108
[10] Gerchikov L, Ipatov A, Solov'yov A and Greiner W 2000 *J. Phys.* B **30** 4905-4926
[11] Surdutovich E and Solov'yov A 2010 *Phys. Rev.* E **82** 051915
[12] Yakubovich A, Surdutovich E and Solov'yov A 2011 *Nucl. Instr. Meth.* B 10.1016/j.nimb.2011.10.069
[13] Barkas W H 1963 *Nuclear Research Emulsions I. Techniques and Theory* vol 1 (New York, London: Academic Press)
[14] Surdutovich E, Obolensky O, Scifoni E, Pshenichnov I, Mishustin I, Solov'yov A and Greiner W 2009 *Eur. Phys. J.* D **51** 63–71
[15] Surdutovich E, Gallagher D and Solov'yov A 2011 *Phys. Rev.* E **84** 051918
[16] Wu B, Mohideen K, Vasudevan D and Davey C 2010 *Structure* **18** 528
[17] Davey C and Richmond T 2002 *Proc. Natl. Acad. Sci. USA* **99** 11169
[18] Phillips J C *et al* 2005 *J. Comp. Chem.* **26** 1781–1802
[19] MacKerell A *et al* 1998 *J. Phys. Chem.* B **102** 3586–3616
[20] Yakubovich A, Surdutovich E and Solov'yov A 2011 *Proceedings of the 7th International Conference (ICAMDATA 2010)* vol **1344** p 230
[21] Range K, McGrath M and York X L 2004 *J. Am. Chem. Soc.* **126** 1654

Role of the interaction processes in the depth-dose distribution of proton beams in liquid water

Rafael Garcia-Molina[1], Isabel Abril[2], Pablo de Vera[2], Ioanna Kyriakou[3] and Dimitris Emfietzoglou[3]

[1] Departamento de Física – Centro de Investigación en Óptica y Nanofísica, Universidad de Murcia, E-30100 Murcia, Spain
[2] Departament de Física Aplicada, Universitat d'Alacant, E-03080 Alacant, Spain
[3] Medical Physics Laboratory, University of Ioannina Medical School, GR-45110 Ioannina, Greece

E-mail: rgm@um.es

Abstract.
We use a simulation code, based on Molecular Dynamics and Monte Carlo, to investigate the depth-dose profile and lateral radial spreading of swift proton beams in liquid water. The stochastic nature of the projectile-target interaction is accounted for in a detailed manner by including in a consistent way fluctuations in both the energy loss due to inelastic collisions and the angular deflection from multiple elastic scattering. Depth-variation of the projectile charge-state as it slows down into the target, due to electron capture and loss processes, is also considered. By selectively switching on/off these stochastic processes in the simulation, we evaluate the contribution of each one of them to the Bragg curve. Our simulations show that the inclusion of the energy-loss straggling sizeably affects the width of the Bragg peak, whose position is mainly determined by the stopping power. The lateral spread of the beam as a function of the depth in the target is also examined.

1. Introduction

The advantages of ion beams in cancer treatment with respect to conventional radiotherapy with photons and electrons are well known [1], due to their characteristic Bragg peak, small lateral spreading and increased relative radiobiological effectiveness (RBE). Nowadays hadrontherapy is a powerful emergent technique used in cancer therapy, especially for deep-seated tumours [2, 3]. Nonetheless, a complete understanding of the relevant involved interactions has not yet been reached, since ion beam radiotherapy involves mechanisms in several spatial and temporal scales [4]. Actually, the whole hadrontherapy problem implies processes from the scale of zeptoseconds and picometers (nuclear fragmentations), passing through femptoseconds and nanometers (primary ion and secondary particles propagation) until millimeters and minutes or even years (tumor death and medical treatments) [5].

In this work we are interested in the first stage of the process, i.e., the depth-dose distribution by the primary ion beam in biological media, since data obtained in this step are needed to model and to understand the subsequent processes. With this aim, we use the Molecular Dynamics and Monte Carlo based code SEICS (Simulation of Energetic Ions and Clusters through Solids) [6] to simulate the motion of fast proton beams in liquid water. Since the code allows us to follow the trajectory, energy and charge state of each incident projectile through the stopping media, we are able to determine both the spatial evolution of the ion beam and its depth-dose profile. The SEICS code considers inelastic collisions between the projectile and the target electrons, corresponding to the electronic stopping force, where energy-loss fluctuations due to the stochastic nature of these processes are included through the energy-loss straggling. The elastic scattering of the projectile with target atoms is also incorporated in the code, as well as the electron capture and loss mechanism by the projectile. Each of these processes can be switched on and off in the simulation, providing information on their effect in the energy evolution of the primary beam, and in the spatial distribution of energy deposition, that is, into the Bragg curve. Although the energies used in cancer treatment are typically of hundreds of MeV, we will focus on low energies (several MeV), where differences of the order of micrometers in the depth-dose curves are more visible.

The work is structured as follows: section 2 briefly reviews the dielectric formalism to evaluate the stopping power and the energy-loss straggling of fast protons in liquid water, and the MELF-GOS methodology to calculate the electronic excitation spectrum of liquid water. These stopping magnitudes are input quantities into the simulation code SEICS, whose main characteristics are featured in section 3. The results of the simulations are presented in section 4, and the final conclusions are given in section 5.

2. Inelastic energy-loss of swift projectiles

A swift charged particle that is travelling through matter experiences Coulomb interactions with target electrons, resulting in electronic excitations and ionizations. As a consequence, the swift particle transfers energy and momentum to the target, resulting in a slowing down, which depends on its velocity. As the projectile can also lose or capture electrons, it feels a stopping force which is strongly dependent on its charge state.

The dielectric formalism [7, 8] provides the framework to calculate the electronic stopping magnitudes taking into account condensed-phase effects of the irradiated target. For a projectile with atomic number z, mass m, and charge-state q travelling with velocity v through a medium with dielectric function $\epsilon(k,\omega)$, the dielectric formalism states that the statistical moments of the energy loss distribution, $\mathcal{F}_q(n)$, are given by

$$\mathcal{F}_q(n) = \frac{2e^2\hbar^{n-1}}{\pi v^2} \int_0^\infty \frac{\mathrm{d}k}{k} \left[z - \rho_q(k)\right]^2 \int_0^{kv} \mathrm{d}\omega \, (\omega)^n \, \mathrm{Im}\left[\frac{-1}{\epsilon\,(k,\omega)}\right] , \tag{1}$$

where $\mathcal{F}_q(1) = S_q$ is the stopping power (mean energy lost per unit path length) and $\mathcal{F}_q(2) = \Omega_q^2$ is the energy-loss straggling (related to the variance of the energy-loss distribution) [9]. Here e is the elemental charge, and $\rho_q(k)$ is the Fourier transform of the projectile electronic density for the charge-state q, which is calculated according to the modified Brandt-Kitagawa model [10, 11]. In this expression, the electronic structure of the target is accounted for by the so-called energy-loss function (ELF), $\mathrm{Im}[-1/\epsilon(k,\omega)]$, which gives the probability of producing an excitation with momentum transfer $\hbar k$ and energy transfer $\hbar\omega$.

Since the electronic stopping magnitudes are strongly affected by the charge-state of the projectile, they must be calculated as a weighted sum over its charge-state fractions

$$\mathcal{F}(n) = \sum_{q=0}^{z} \phi_q \, \mathcal{F}_q(n) , \tag{2}$$

with ϕ_q being the probability of finding the projectile with a charge-state q, which depends on the target as well as on the nature and velocity of the projectile. As the dynamical equilibrium is quickly reached, we use the equilibrium charge fractions from a parameterization to experimental data [12].

The target-dependent magnitude in equation (1) is the ELF. Calculating the excitation spectrum from first principles for condensed phases requires a vast computational effort, and therefore, it is preferable to use semiempirical approaches such as the optical-data models, which are computationaly more feasible and give fairly good results [13, 14]. In these optical-data models, the ELF in the optical limit ($k = 0$) is taken from experimental data and then extended to $k \neq 0$ by suitable dispersion schemes, providing in this manner the whole Bethe surface.

Here the Bethe surface of liquid water is obtained from optical data [15] through the MELF-GOS methodology (Mermin Energy-Loss Function – Generalized Oscillator Strengths). It is based on the separation of the ELF in outer and inner electronic excitation contributions, namely [16, 17],

$$\mathrm{Im}\left[\frac{-1}{\epsilon(k,\omega)}\right] = \mathrm{Im}\left[\frac{-1}{\epsilon(k,\omega)}\right]_{\mathrm{outer}} + \mathrm{Im}\left[\frac{-1}{\epsilon(k,\omega)}\right]_{\mathrm{inner}} . \tag{3}$$

As the inner-shell electrons preserve their atomic character, they can be properly described by their generalized oscillator strengths (GOS) in the hydrogenic approach. Aggregation effects are contained in the outer-electrons excitation contribution to the ELF, which we describe through a weighted sum of Mermin-type ELF. The Mermin dielectric function [18] improved the Lindhard dielectric function for a free electron gas [7], since the damping mechanism of plasmons was introduced phenomenologically, resulting in a more realistic description of the ELF, where the broadening of the excitations is taken into account. Since the Mermin dielectric function has an analytic dependence on the momentum transfer, $\hbar k$, it has the advantage that an extension algorithm to $k \neq 0$ is not necessary, as it is done in other models [14, 19, 20]. A detailed explanation of the MELF-GOS model can be found elsewhere [17].

The most recent set of experimental data for the ELF of liquid water was provided by the Sendai group [15, 21, 22] by using inelastic X-ray scattering spectroscopy (IXSS) to measure the generalized oscillator strength (GOS) of liquid water at several momentum transfers. The IXSS data at the optical limit extend from 6 to 160 eV excitation energies, providing a near complete knowledge of the dielectric response properties of the valence-shells of liquid water. Therefore, we account for the outer electrons contribution to the ELF (at $k = 0$) by fitting the experimental ELF using a sum of Mermin-type ELF, $\mathrm{Im}[-1/\epsilon_{\mathrm{M}}(k,\omega)]$,

$$\mathrm{Im}\left[\frac{-1}{\epsilon(k=0,\omega)}\right]_{\mathrm{outer}} = \mathrm{Im}\left[\frac{-1}{\epsilon(k\approx 0,\omega)}\right]_{\mathrm{exp}} = \sum_i \frac{A_i}{W_i^2}\mathrm{Im}\left[\frac{-1}{\epsilon_{\mathrm{M}}(k=0,\omega;W_i,\gamma_i)}\right]\Theta(\omega-\omega_{\mathrm{th},i}) , \tag{4}$$

where A_i, W_i and γ_i are, respectively, the intensity, position and width of each Mermin-type ELF, and $\omega_{\mathrm{th},i}$ is a threshold energy.

The Bethe surface measurements [21, 22] of liquid water by IXSS at several values of the momentum transfer $\hbar k$ are shown by symbols in figure 1 together with the results of the MELF-GOS model (red solid lines). As can be seen, the Bethe surface obtained by the MELF-GOS method is in very good agreement with experimental data in a wide range of k-values. We want to stress that the choice of the extension algorithm in the ELF is crucial to properly reproduce the Bethe surface. A discussion of the influence of the different methods used to extend the optical ELF of liquid water at non-zero momentum transfer can be found in references [14, 20].

The stopping power and the energy-loss straggling of protons in liquid water have been calculated using the dielectric formalism (equations (1) and (2)) and the MELF-GOS model described previously. The results are shown in figure 2 (black solid lines) as a function of the proton energy and compared with several experimental data for liquid [23, 24, 25] and solid

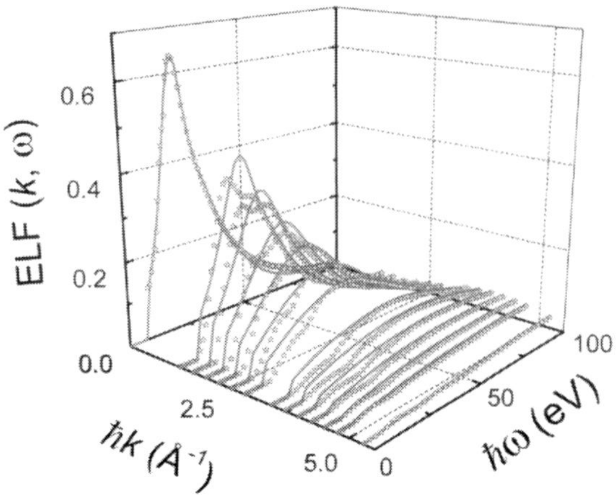

Figure 1. Bethe surface of liquid water. Symbols represent experimental data for several values of the momentum transfer $\hbar k$ measured by the Sendai group [21, 22], whereas solid lines correspond to the MELF-GOS model.

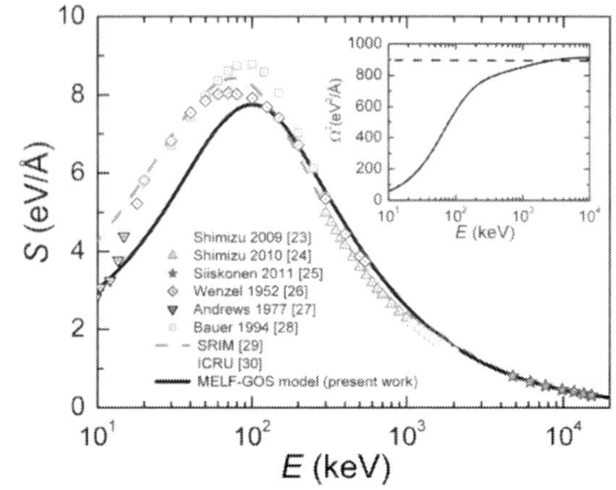

Figure 2. Stopping power S (and energy-loss straggling Ω^2, inset) of liquid water for a proton beam, as a function of its energy. Symbols represent experimental data whereas black solid line corresponds to the MELF-GOS model. See the text for more details.

[26, 27, 28] water (symbols), as well as with semi-empirical calculations such as SRIM [29] (red dashed line) and ICRU recommended data [30] (green dotted line). All the predictions agree quite well at high energies, among them that with the most recent experimental data for liquid water [25], but differences appear at the region of the maximum stopping power. It is worth to note that ICRU and SRIM data are parameterized to ice data, the only ones available at the maximum stopping region. The dielectric calculations with the MELF-GOS model yield lower stopping power values at energies around and below the maximum stopping, which could be attributed to chemical and phase effects, since this model describes self-consistently the electronic excitations of liquid water in a realistic approach, as it has been shown in figure 1.

Stopping power calculations on the depth-dose profile are significant because they mainly determine the most important features of the Bragg curve. Therefore, uncertainties in the stopping power directly produce (undesired) ambiguities in radiotherapy dosimetry. The inset of figure 2 shows the energy-loss straggling of protons in liquid water, as a function of their energy, calculated with the dielectric formalism and the MELF-GOS model. As expected, it tends at high energies to the Bohr straggling limit of a free-electron target, which is depicted by a horizontal dashed line in the figure.

3. Simulation of swift particles moving through liquid water

We have used the SEICS code [6, 31] to simulate the trajectories of fast protons in liquid water, which permits to obtain their spatial and energy distributions at any depth. SEICS is based in Molecular Dynamics and Monte Carlo techniques to follow in detail the motion of a charged projectile through a stopping medium. In each step of the simulation, defined by a time step Δt, the equation of motion of the projectile is numerically solved by the velocity variant of Verlet's algorithm [32] in order to compute its new position $\vec{r}$ and velocity $\vec{v}$

$$\vec{r}(t+\Delta t) = \vec{r}(t) + \vec{v}(t)\Delta t + \frac{\vec{F}(t)}{2m}\,(\Delta t)^2\,\left[1-(v(t)/c)^2\right]^{3/2}\,, \tag{5}$$

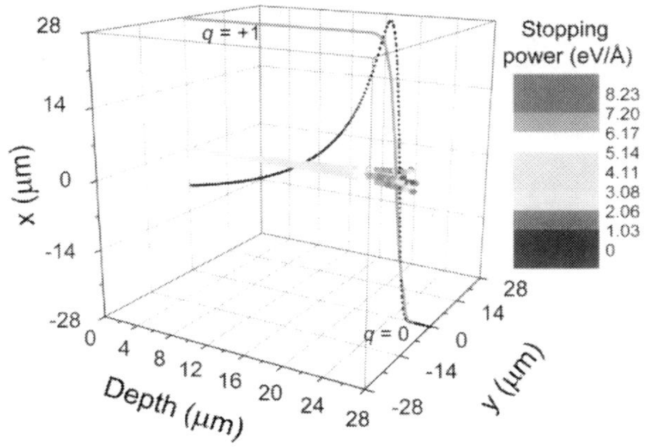

Figure 3. Simulated trajectories corresponding to 1 MeV proton beam incident in liquid water. The graded color along the trajectories represent the stopping power at each depth. The depth-dose curve is shown qualitatively by a dotted black line, and the average charge-state of the projectile by a solid magenta line.

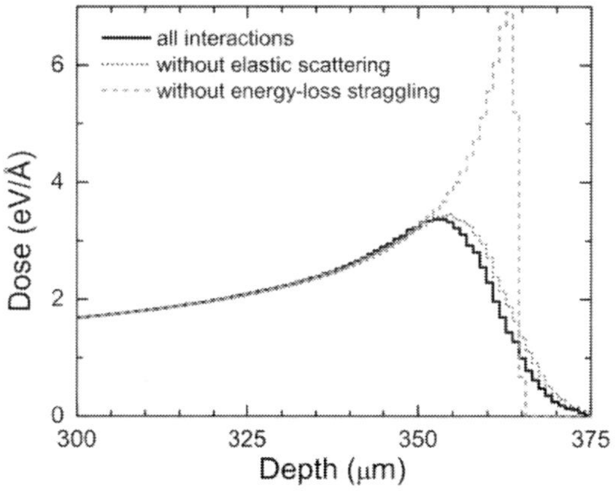

Figure 4. Depth-dose distribution for a 5 MeV proton beam in liquid water. Solid curve represents the result obtained when all the interactions are considered, whereas the dashed and dotted curves where obtained by removing different interactions in the simulation code SEICS.

$$\vec{v}(t+\Delta t) = \vec{v}(t) + \frac{\vec{F}(t)+\vec{F}(t+\Delta t)}{2m}\,\Delta t\,\left[1-(v(t)/c)^2\right]^{3/2} . \tag{6}$$

The factor in brackets is a correction to account for relativistic velocities of the projectiles.

The electronic stopping force $\vec{F}$ that the projectile experiences is obtained from the statistical moments of the energy loss distribution, which were calculated in the previous section with the dielectric formalism and the MELF-GOS model. The stochastic nature of the electronic interactions leads to an electronic stopping force described by a Gaussian distribution centered in the stopping power

$$\vec{F} = -\left[S_q + (\Omega_q/\sqrt{\Delta s})\sqrt{-2\ln\xi_1}\cos(2\pi\xi_2)\right]\hat{v} , \tag{7}$$

where $\hat{v}$ is the unit vector of the instantaneous projectile velocity $\vec{v}$; S_q and Ω_q^2 are, respectively, the stopping power and the energy-loss straggling of liquid water for a projectile with charge q, obtained from equation (1), and $\Delta s = v\Delta t$ is the distance travelled by the projectile in a time step Δt. The symbols ξ_1 and ξ_2 refer to random numbers uniformly distributed between 0 and 1.

To speed up the simulation, at higher projectile energies ($E \geq 10$ MeV/amu) we use the Bohr energy-loss straggling and the Bethe-Bloch stopping power formula [33], with a mean excitation energy $I = 79.4$ eV for liquid water obtained from the MELF-GOS model [31].

The simulation code SEICS also takes into account multiple elastic scattering between the projectile and the target nuclei, by a Monte Carlo algorithm [34, 35] using the universal interatomic potential with a screening length [29]; for more details on the SEICS implementation see reference [6]. Elastic interactions are responsible for the projectile angular deflection and, at the end of their travel, nuclear energy-loss starts to be considerable.

The charge exchange of the projectile due to electron capture and loss processes through the target is also considered in SEICS by a Monte Carlo procedure [6]. In this manner, the charge

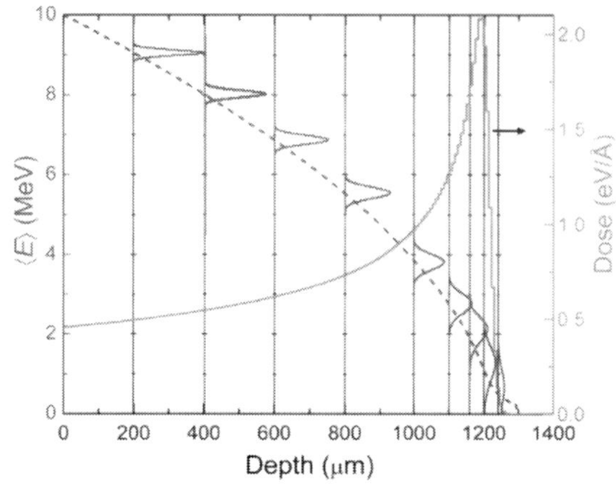

Figure 5. (Left axis) Mean energy (blue dashed line) for a 10 MeV proton-beam incident in liquid water, as a function of its depth. The projectile energy-distributions are depicted by blue solid histograms at several depths. (Right axis) The depth-dose distribution is shown by a grey line.

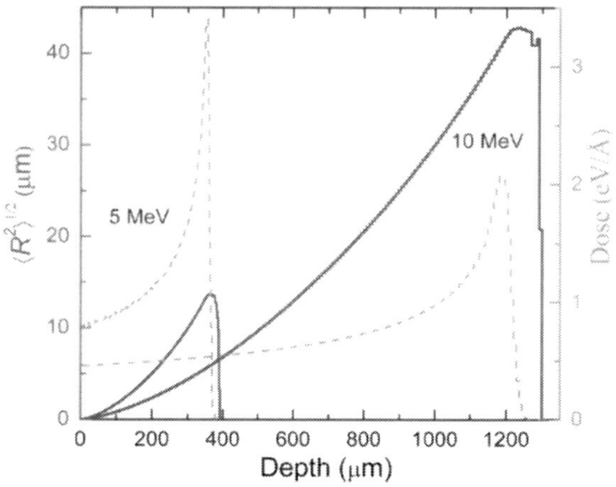

Figure 6. (Left axis) Root mean square radius of the beam (blue solid lines) for 5 MeV and 10 MeV proton beams incident in liquid water, as a function of the depth. (Right axis) The corresponding depth-dose distributions (grey dashed lines) are shown for comparison.

states of the projectile are known along its track, and the stopping power and the energy-loss straggling are calculated at each depth according to the actual charge state of the projectile.

4. Results and discussion

We present the results obtained from the simulation with the SEICS code of the irradiation of liquid water with protons. The main input quantities to be used in the code are the electronic stopping power and the energy-loss straggling, which were obtained in section 2.

Figure 3 depicts several trajectories of 1 MeV protons in liquid water, showing qualitatively the capabilities of the SEICS code, where the graded colors represent the stopping power at each depth. It illustrates (by graded colors) the increase in the dose delivered by the protons along their paths, reaching a narrow maximum at the end of their trajectories (the Bragg peak), where the beam has spread out due to multiple elastic scattering. Although $\sim 10^4$ histories are needed to obtain good statistical results, only a few trajectories are displayed for clarity. The depth dose distribution is depicted by a dotted black line, which clearly represents the Bragg peak. The projectile average charge (magenta solid line) is shown as a function of the depth; note that the mean charge state is +1 in almost all the track, and it rapidly drops to 0 at the end of the Bragg peak. Therefore, at the distal part of the Bragg peak, the number of neutral hydrogen atoms overcomes the number of protons, a fact that is of relevance to determine the secondary electron spectra induced by the primary projectile, which is of most importance to determine the cellular damage at the molecular level.

The simulated depth-dose distribution (i.e. the Bragg curve) of a 5 MeV proton beam in liquid water is shown in figure 4. The black solid line represents the full simulation when elastic scattering and energy-loss straggling are both included in the calculations. By switching on/off the fluctuations in the inelastic energy loss (through the energy-loss straggling) and in the elastic scattering we study how these processes affect the depth-dose profile. The red dashed line represents the depth-dose curve obtained when removing the energy-loss straggling, whereas the blue dotted line is the simulation without the elastic scattering. As can be seen, both processes practically have no effect on the plateau of the depth-dose curve. Nonetheless, removing the energy-loss straggling does affect the maximum dose region by shifting the Bragg peak to larger depths and strongly narrowing its shape. Removing the elastic scattering has a less important effect, shifting the distal part of the Bragg curve deeper in the target, but not changing its shape. Therefore, the results depicted in figure 4 clearly indicate that the elastic scattering and mainly the energy-loss straggling are essential to determine the right position and shape of the Bragg peak.

The mean energy of the projectile as a function of the depth (blue dashed line) corresponding to a 10 MeV proton beam in liquid water is shown in figure 5. For comparison, the depth-dose curve is shown as a grey solid line. Also, the projectile energy distributions at several depths are depicted by blue (little) histograms, which exhibit a broadening due to the fluctuations in the inelastic collisions accounted for the energy-loss straggling; the reduction of the number of particles as they are stopped is responsible of the smaller size and asymmetry of the deeper histograms. At the Bragg peak the proton beam cannot be considered monoenergetic, so its energy distribution must be carefully taken into account since it is strongly correlated with the spectrum of generated secondary electrons.

Finally, we analyze the lateral spreading of the proton beam in liquid water due to multiple elastic scattering. The root mean square radius of the beam, $\langle R^2 \rangle^{1/2}$, is depicted by solid blue lines in figure 6 for 5 MeV and a 10 MeV proton beams as a function of the depth in liquid water. The corresponding depth-dose curves (grey dashed lines) are also shown for comparison. It is observed how, due to elastic collisions, the mean square radius of the beam increases with the depth following a parabolic dependence, falling off at the distal region of the Bragg curve. This fact is explained by taking into account that only very few projectiles travel in almost a straight line to reach these deep regions. Therefore, as most of the projectiles deviate from their initial direction, they are stopped at lower depths.

5. Conclusions

The depth-dose profile of swift proton beams in liquid water has been simulated by the SEICS code, which is based on Molecular Dynamics and Monte Carlo techniques, allowing a detailed monitoring of the motion of charged particles through condensed media. The simulation includes the following interactions: electronic interaction (evaluated from the stopping power and its fluctuations due to energy-loss straggling), multiple elastic scattering and charge exchange between the projectile and the target due to electron capture and loss processes. A realistic electronic excitation spectrum of liquid water has been used through the MELF-GOS optical-data model, to account for its particular excitation spectrum.

We have evaluated the influence of each interaction in the depth-dose curves of proton beams in liquid water, concluding that fluctuations in the energy-loss processes (through the energy-loss straggling) are fundamental to determine the shape and position of the Bragg peak. Multiple elastic scattering has a minimum effect on the depth-dose distribution, changing the range of the protons only at the distal part of the Bragg curve, nonetheless it is the main responsible of the broadening of the beam profile.

From our results we conclude that, due to the energy-loss straggling, the beam cannot be considered as monoenergetic at the Bragg peak, and that the lateral broadening of the beam increases rather rapidly (with a parabolic dependence) as a function of the depth. These features of the proton beam at the Bragg peak are crucial to describe the spectrum of induced secondary electrons in order to do further and more complex simulations for hadronterapy purposes.

Acknowledgments
This work has been financially supported by the the Spanish Ministerio de Ciencia e Innovación (Project FIS2010-17225) and the European Union FP7 ANTICARB (HEALTH-F2-2008-201587). PdV thanks the Conselleria d'Educació, Formació i Ocupació de la Generalitat Valenciana for its support under the VALi+d program. We acknowledge support provided by the COST Action MP 1002, Nanoscale Insights into Ion Beam Cancer Therapy.

References

[1] Nikjoo H, Uehara S, Emfietzoglou D and Brahme A 2008 *New. J. Phys.* **10** 075006
[2] Kraft G 2000 *Prog. Part. Nucl. Phys.* **45** S473
[3] Goitein M, Lomax A J and Pedroni E 2002 *Phys. Tod.* **55** 45
[4] Solov'yov A V, Surdutovich E, Scifoni E, Mishustin I and Greiner W 2009 *Phys. Rev.* E **79** 011909
[5] Surdutovich E, Scifoni E and Solov'yov A V 2010 *Mutat. Res.* **704** 206
[6] Garcia-Molina R, Abril I, Heredia-Avalos S, Kyriakou I and Emfietzoglou D 2011 *Phys. Med. Biol.* **56** 6475
[7] Lindhard J 1954 *Dan. Vid. Selsk. Mat. Fys. Medd.* **28** No 1
[8] Ritchie R H 1957 *Phys. Rev.* **106** 874
[9] Sigmund P 2006 *Particle Penetration and Radiation Effects. General Aspects and Stopping of Swift Point Charges* (Springer Series in Solid-State Sciences Vol. 151) (Berlin: Springer)
[10] Brandt W and Kitagawa M 1982 *Phys. Rev.* B **25** 5631
[11] Brandt W 1982 *Nucl. Instr. Meth.* **194** 13
[12] Schiwietz G and Grande P L 2001 *Nucl. Instr. Meth.* B **175-177** 125
[13] Ritchie R H and Howie A 1977 *Philos. Mag.* **36** 463
[14] Garcia-Molina R, Abril I, Kyriakou I and Emfietzoglou D 2012 Energy loss of swift protons in liquid water: Role of optical data input and extension algorithms, Ch. 15 in *Radiation Damage in Biomolecular Systems*, eds G García Gómez-Tejedor and M C Fuss (Dordrecht: Springer)
[15] Hayashi H, Watanabe N, Udagawa Y and Kao C C 2000 *Proc. Nat. Acad. Sc. USA* **97** 6264
[16] Abril I, Garcia-Molina R, Denton C D, Pérez-Pérez F J and Arista N R 1998 *Phys. Rev.* A **58** 357
[17] Heredia-Avalos S, Garcia-Molina R, Fernández-Varea J M and Abril I 2005 *Phys. Rev.* A **72** 052902
[18] Mermin N D 1970 *Phys. Rev.* **1** 2362
[19] Emfietzoglou D, Garcia-Molina R, Kyriakou I, Abril I and Nikjoo H 2009 *Phys. Med. Biol.* **54** 3451
[20] Emfietzoglou D, Kyriakou I, Abril I, Garcia-Molina R and Nikjoo H 2012 *Int. J. Radiat. Biol.* **88** 22
[21] Watanabe N, Hayashi H and Udagawa Y 1997 *Bull. Chem. Soc. Jap.* **70** 719
[22] Watanabe N, Hayashi H and Udagawa Y 2000 *J. Phys. Chem. Sol.* **61** 407
[23] Shimizu M, Kaneda M, Hayakawa T, Tsuchida H and Itoh A 2009 *Nucl. Instr. Meth.* B **267** 2667
[24] Shimizu M, Hayakawa T, Kaneda M, Tsuchida H and Itoh A 2010 *Vacuum* **84** 1002
[25] Siiskonen T, Kettunen H, Peräjärvi K, Javanainen A, Rossi M, Trzaska W H, Turunen J and Virtanen A 2011 *Phys. Med. Biol.* **56** 2367
[26] Wenzel W A and Whaling W 1952 *Phys. Rev.* **87** 499
[27] Andrews D A and Newton G 1977 *J. Phys. D: Appl. Phys.* **10** 845
[28] Bauer P, Kaferbock W and Necas V 1993 *Nucl. Instr. Meth.* B **93** 132
[29] Ziegler J F, Biersak J P and Ziegler M D 2008 *SRIM. The Stopping and Range of Ions in Matter* (Chester, MD: SRIM Co). `http://www.srim.org`
[30] International Commission on Radiation Units and Measurements 1993 *Stopping Powers and Ranges for Protons and Alpha Particles* (ICRU Report 49) (Bethesda, MD: ICRU)
[31] Garcia-Molina R, Abril I, Denton C D, Heredia-Avalos S, Kyriakou I and Emfietzoglou D 2009 *Nucl. Instr. Meth.* B **267** 2647
[32] Allen M P and Tildesley D J 1989 *Computer simulation of liquids* (Oxford: Oxford University Press)
[33] Inokuti M 1971 *Rev. Mod. Phys.* **43** 297
[34] Möller W, Pospiech G and Schrieder G 1975 *Nucl. Instr. Meth.* **130** 265
[35] Zajfman D, Both G, Kanter E P and Vager Z 1990 *Phys. Rev.* A **41** 2482

Journal of Physics: Conference Series **373** (2012) 012016 doi:10.1088/1742-6596/373/1/012016

Relativistic protons for image-guided stereotactic radiosurgery

M Durante and H Stöcker

GSI Helmholtz Center for Heavy Ion Research, Planckstrasse 1, Darmstadt, Germany; and Frankfurt Institute for Advanced Studies (FIAS), Frankfurt, Germany

E-mail: M.Durante@gsi.de

Abstract. Bragg-peak radiosurgery and proton radiography have been used in radiotherapy over the past few years. Non-Bragg-peak (plateau) relativistic protons (E>1 GeV) can offer advantages both in terms of precision and target margin reduction, and especially thanks to the possible simultaneous use of high-resolution online proton radiography. Here we will present initial simulations and experiments toward image-guided stereotactic radiosurgery using GeV protons.

1. Introduction

Proton therapy is now a well established method in the treatment of cancer [1] and noncancer diseases [2]. The rationale of using protons with energies between 60 and 250 MeV is based on the favourable depth-dose distribution, so that the targets can be located on a spread-out Bragg peak while the normal tissue is exposed on the plateau region. However, the proton beam is broadened by multiple scattering in the beamline materials and in the patient's body. This broadening produces a "dose halo" in the treatment plan and worsens the dose contours [3]. For a charged particle with atomic number z and mass number A, the lateral scattering is roughly proportional to $z/A\beta^2$ (where β is the ratio of the particle velocity and the speed of light c), and therefore, the lateral scattering can be reduced either using heavy ions such as carbon [4], or by increasing the particle velocity. In the latter case, using protons in the GeV region the targets cannot be exposed on the Bragg peak (for instance, the range of 1 GeV protons is ~3.2 m in water), and then requires various beams cross-fired to the target from different angles, similarly to X-ray therapy.

The advantage of the "plateau" (non-Bragg-peak) radiotherapy will be a stable beam profile providing very sharp dose contours for sparing of critical organs. For this reason, relativistic protons were proposed for plateau stereotactic radiosurgery already in Berkeley [5]. The only clinical experience comes from St. Petersburg in Russia: more than 1,000 patients have been treated with 1 GeV protons at the Petersburg Nuclear Physics Institute (PNPI) since 1975, and results were reported in the present journal [6].

Even though the low scattering is appealing, the loss of the favourable dose-depth distribution provided by the Bragg curve is of course a major drawback compared to Bragg-peak radiotherapy. However, an additional advantage of relativistic protons is that the beam crossing the patient can be exploited for proton radiography. Proton radiography was investigated since the early 1970s because of its low radiation dose and high density resolution, but until recently the image blurring due to multiple Coulomb scattering was limiting its practical applications [7]. About a decade ago researchers of Los Alamos National Laboratory (LANL) have introduced a magnetic lens after the object for imaging and chromatic aberration corrections. This allowed to reach unprecedented spatial resolution with high-energy protons, as proved by many dynamic experiments with 800 MeV beams at LANL [8]. A project for proton microscopy (PRIOR) at the new Facility for Anti-proton and Ion Research (FAIR) in Darmstadt plans to exploit a 4.5 GeV proton beam for radiography, reaching a spatial resolutions below 10 μm and a time resolution below 10 ns [9]. The high-precision in beam delivery combined with online high-resolution imaging and dose verification leads to reduced target margins and improved image-guided stereotactic proton radiosurgery (IGSpRS) for cancer (e.g. small brain metastasis, pituitary adenoma, vestibular Schwannoma) and noncancer (e.g. arteriovenous malformations, trigeminal neuralgia, epilepsy, intracranial aneurysm, macular degeneration) lesions [2] (figure 1).

Figure 1. Proposed setup for IGSpRS using relativistic protons. Patient has to be rotated to allow cross-firing of the proton beam from different angles. A magnetic lens system is used for high-resolution radiography. This imaging system makes possible online guidance.

2. Lateral scattering

One advantage of plateau compared to Bragg-peak protons is the reduced lateral scattering. Dose halo due to scattering is indeed a major hindrance for treatment of small targets close to critical organs in protontherapy.

Proton scattering in treatment plans is generally well described by the Molière theory [10]. For small angles, the higher order terms in the Molière theory can be neglected and the beam profile at a depth d into a material of radiation length L_{rad} can be described as a Gaussian with standard deviation (p = particle momentum):

$$\sigma(rad) = 14.MeV \frac{z}{\beta pc}\sqrt{\frac{d}{L_{rad}}}\left[1+\frac{1}{9}\log_{10}\left(\frac{d}{L_{rad}}\right)\right] \quad (1)$$

Figure 2 shows a beam's eye view of two protons beams going through 15 cm of water. The Monte Carlo simulation by SRIM2011 calculated the xy plane coordinates (in cm) of single protons with initial energy of 1 GeV (residual range in water 322 cm) or 150 MeV (residual range in water 15.6 cm) shot in the (0,0) central position. Clearly, relativistic protons in plateau (blue dots) are very sharp compared to the broad scattering of Bragg-peak protons (red dots).

The reduced lateral scattering is clearly shown in figure 3, where the beam FWHW in water was calculated using either equation (1) or the GEANT4 Monte Carlo code [11]. Beams at energies from 60 to 2500 MeV are simulated. Protontherapy of eye's tumor requires 60 MeV, while deep

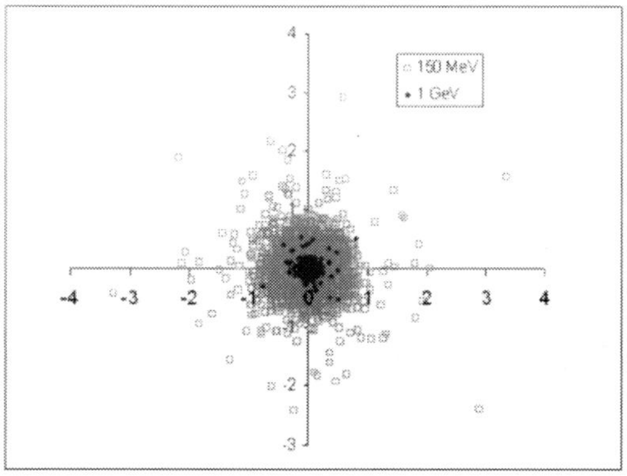

Figure 2. Lateral scattering of proton beams at two different energies after 15 cm in water. Each dot represents a single proton. Simulation by SRIM2011

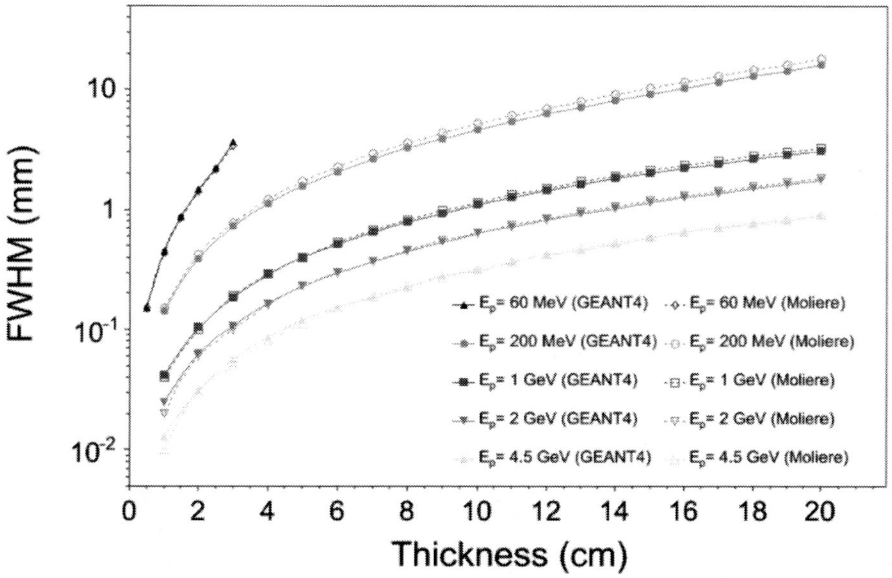

Figure 3. Simulation of the FWHM of proton beams in water at different energies using either a deterministic or a Monte Carlo calculation. Molière equation is shown in equation (1).

protontherapy of solid tumors exploits beams between 150 and 250 MeV. Energies close to 1 GeV (around 800 MeV) are available at Los Alamos National Laboratory (LANL) in USA and ITEP in Russia for proton radiography. These energies can also be reached by medical synchrotrons used to accelerate C-ions up to 400 MeV/n, such as those at HIT in Heidelberg (Germany) and CNAO in Pavia (Italy). Beams of 1-2 GeV are available at several high-energy accelerators with biomedical applications, such as the NSRL at the Brookhaven National Laboratory in USA and SIS-18 at GSI in Germany. Finally, the 4.5 GeV beam has been proposed for proton microscopy at FAIR, the facility under construction in Darmstadt that will exploit the SIS-18 as injector [12].

3. Depth-dose distribution

Although GeV protons must be used in the "plateau" region of the Bragg curve, the depth-dose distribution is not flat. This is caused by the production of secondary protons by nuclear reactions, particularly knock-out and evaporation hadrons. These effects have been carefully studied in the framework of shielding of relativistic protons for space radiation protection [13]. As shown in the Monte Carlo simulation in figure 4, the effect is important, and leads to a 30-40% increase of the dose with depth in a thick (~20 cm) tissue.

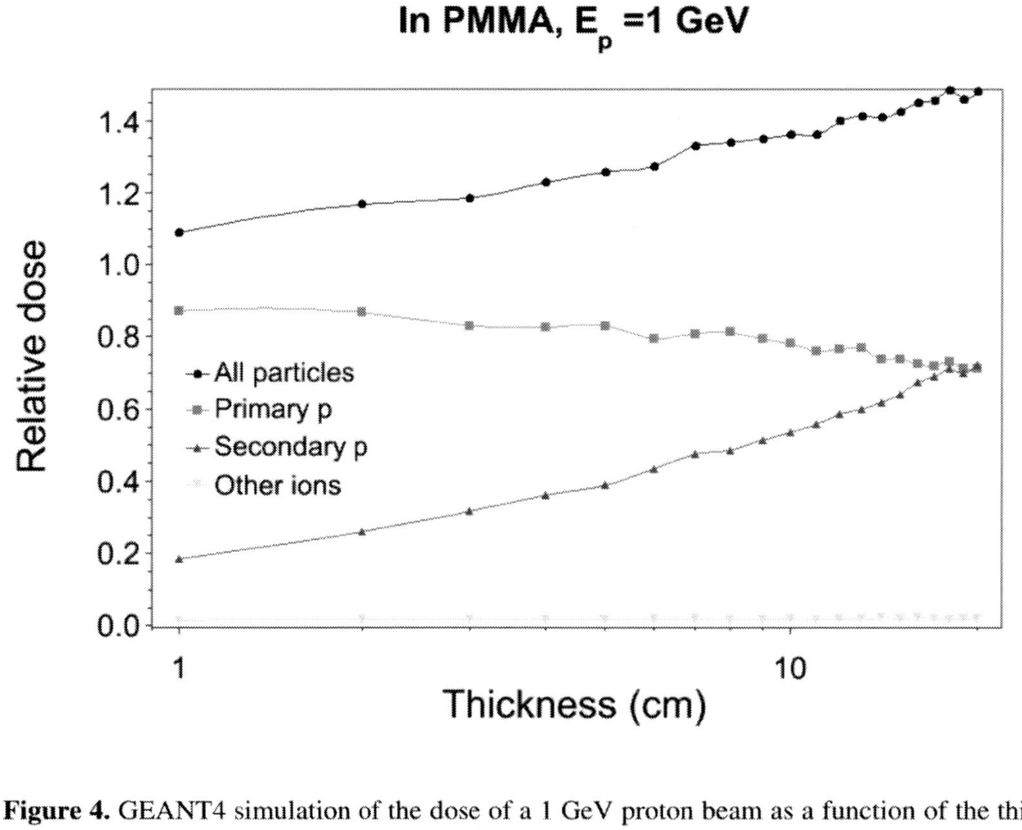

Figure 4. GEANT4 simulation of the dose of a 1 GeV proton beam as a function of the thickness in a plastic (Lucite, PMMA) phantom. Dose is normalized to the entrance value. The contributions of primary protons, secondary protons, and other ions (neutrons, deuterons, tritons, helium, and lithium) are plotted in different colours and symbols.

Journal of Physics: Conference Series **373** (2012) 012016 doi:10.1088/1742-6596/373/1/012016

The increase as a function of the depth is mostly caused by fast knock-out protons. These results and simulations show that in the treatment planning for IGSpRS a careful calculation of the nuclear interaction at relativistic energies is necessary. This will be done within the analytical and Monte Carlo methods developed at the Frankfurt Institute for Advanced Studies (FIAS) [14].

4. Proton radiography

The original idea [15] of proton radiography was to exploit the different ranges of protons crossing structures with different mass thickness, and is indeed now used for quality control in protontherapy [16]. However, a strong image blur is caused by multiple Coulomb scattering. Morris and colleagues in LANL had the idea to compensate the imaging blurring using magnetic lenses (figure 5). Using high-energy protons, optical aberrations are greatly reduced, since both the object scattering and the detector blur are proportional to p^{-1}. The excellent results at LANL [8] for spatial resolution and mass thickness (areal density) measurements led to further plans for high-resolution proton radiography, including the PRIOR at FAIR [9], where 4.5 GeV protons will be used. The PRIOR theoretical spatial resolution is only 20 μm, thus clearly marking a switch from radiography to microscopy. However, these proposals are mostly focused on shock waves, while in our plans (figure 1) we propose to image biological structures, such as tumors, nerves or blood vessels.

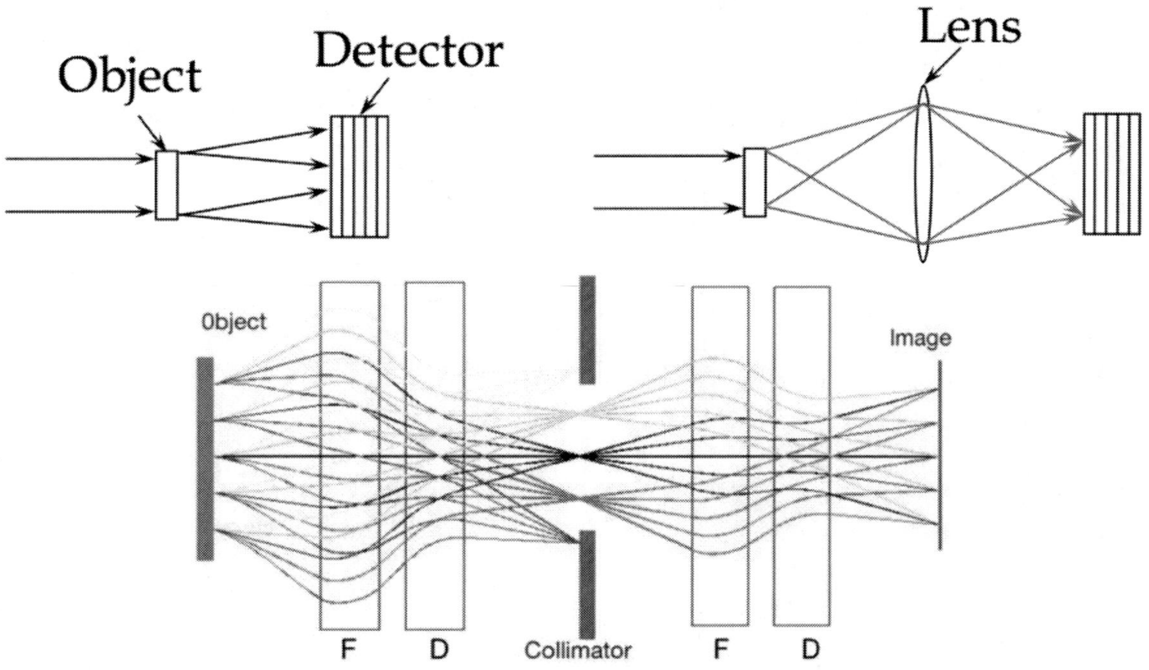

Figure 5. Correction of image blurring in proton radiography using magnetic lenses. The idea was originally proposed by Morris and Zumbro (LANL Technical Report LA-UR-974172, 1997).

The question is therefore whether the novel high-resolution optically-compensated proton radiography can be used for imaging biological objects. A first test has been performed at ITEP in Moscow (Russia), where a proton radiography setup is installed on the 800 MeV beamline. Although the performance was much lower than that expected at PRIOR (e.g. spatial resolution about 0.15 mm), we were able to generate a first image of a zebrafish fixed in paraffin and exposed in a vacuum chamber (figure 6) [17].

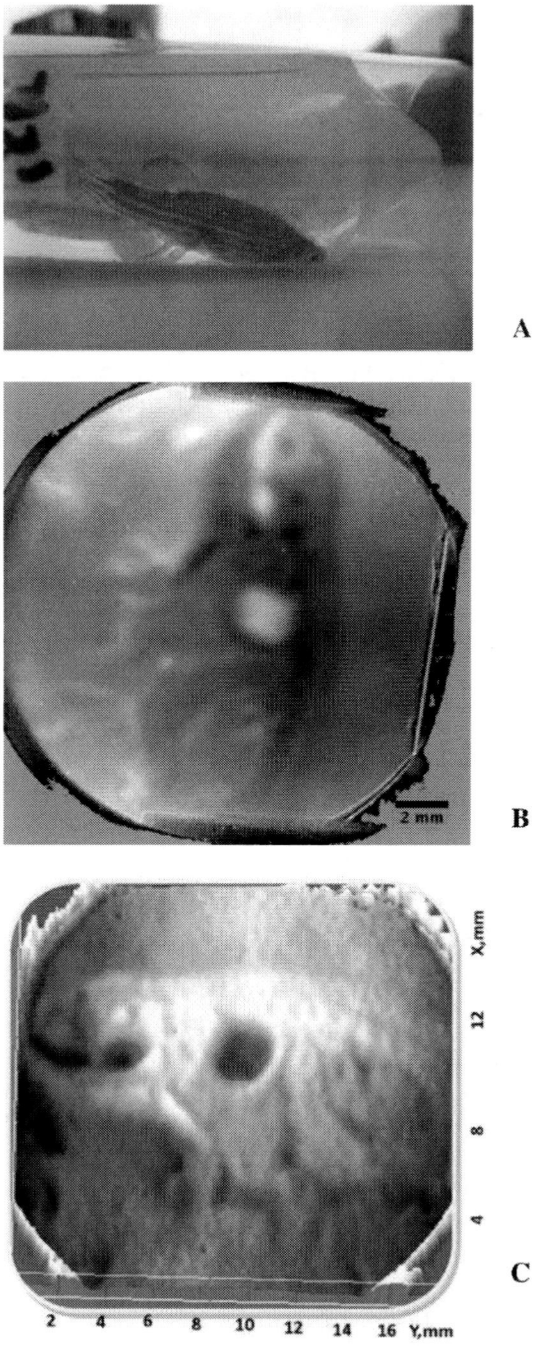

Figure 6. First tests of proton radiography at ITEP, Moscow. A zebrafish (*Danio rerio*) fixed in paraffin (**A**) was exposed in vacuum to 800 MeV protons from the TWAC accelerator using the PUMA-x4 microscope. A 7 mm-diameter, 38 mm long collimator has been used for all the measurements at an intensity of about 1-$3{\cdot}10^{10}$ protons/image. The images at the scatterer (the position right before the matching triplet upstream the microscope) were analyzed. The beam origin and the beam intensity were determined and the shots with largely different beam positions or beam intensities were excluded from the processing (subjective selection). Then, all the remaining images at the detector plane were summed up for each target irradiation sequence, normalized to the shot intensity. While averaging, a 3 or 5-pixel wide median filter has been applied to each individual image to illuminate the dead pixels due to the neutron background. The same procedure has been applied to the white field measurements. All the intermediate image processing steps have been performed storing the data in the 32-bit float TIFF format, in order to avoid information losses. The resulting averaged target images were divided by the corresponding averaged white fields. The 2D image is shown in panel **B**, and a 3D color- reconstruction in panel **C** (rotated 90° with respect to B). The white spot in panel B (in false-purple-color in panel C) corresponds to the fish's empty swim bladder. Details in reference [17].

Images show both the actual 2D imaging accumulated with 100 pulses from the synchrotron of about 10^{10} protons/pulse, and a 3D color reconstruction of the areal density of the sample. Although this should be considered only as a preliminary test, clearly imaging of biological objects is feasible and can provide quantitative information on the areal density of the target [17].

5. Discussion

We presented a proposal for IGSpRS using relativistic protons (figure 1). Thanks to the low lateral scattering (figure 3) and simultaneous online imaging using the high-resolution proton microscopy setup (figure 5), this setup allows reduction of target margins, sparing of critical structures, and target dose escalation, thus potentially leading to improved clinical outcomes. The first step will be the implementation of a treatment plan for IGSpRS, based on the TRiP98 code developed for heavy-ion therapy at GSI [18]. This development will require careful analysis of relativistic nuclear collisions, which contribute significantly to the dose at high energy (figure 6), in a region where medical physics measurements are absent. Mathematical models developed at FIAS [19] will be used for the implementation of relativistic molecular dynamics in the treatment plan. *In silico* trials will then be run comparing IGSpRS to X-ray stereotactic radiosurgery and heavy-ion therapy. In parallel, proton radiography will be tested at GSI and at other accelerators, based on the first encouraging preliminary results (figure 6). Finally, first tests with animal and anthropomorphic phantoms will be performed at GSI in preparation of the final setup. The encouraging clinical results obtained at PNPI with 1 GeV protons [6] without target imaging, support the rationale for IGSpRS for future clinical applications in radiosurgery at FAIR. This prototype could eventually be adopted by other facilities equipped with high-energy proton and heavy ion synchrotrons in clinical environments.

Acknowledgements

We thank Dr. Dmitry Varentsov (GSI) for figure 5, for performing the experiments at ITEP (Moscow, Russia) and for the support of PRIOR to our proposal. For the experiments shown in figure 6, we are also grateful to Dr. Francesco Natale (FIAS) and Dr. Palmina Simoniello (GSI) for the preparation of the biological samples and to the ITEP accelerator crew for the excellent assistance. We also thank Dr. Marie Vanstalle (GSI) for the GEANT4 simulations, Dr. Katia Parodi (HIT, Heidelberg) for useful discussions, and Dr. Zhan Yu (Fudan Cancer Center, Shanghai) for the cartoon in figure 1.

References

[1] Durante M and Loeffler J S 2010 *Nat. Rev. Clin. Oncol.* **7** 37
[2] Bert C, Engenhart-Cabillic R and Durante M 2012 *Med. Phys.* **39** 1716
[3] Schwaab J, Brons S, Fieres J and Parodi K 2011 *Phys. Med. Biol.* **56** 7813
[4] Schardt D, Elsässer T and Schulz-Ertner D 2010 *Rev. Mod. Phys.* **82** 383
[5] Tobias C A 1973 *Radiology* **108** 145
[6] Abrosimov N K, Gavrikov Y A, Ivanov E M, *et al.* 2006 *J. Phys.* **41** 424
[7] Schipper J M and Lomax A J 2011 *Acta Oncol.* **50** 838
[8] Rigg P A, Schwartz C L, Hixson R S, *et al.* 2008 *Phys. Rev.* **B77** 220101
[9] Merrill F E, Golubev A A, Mariam F G, Turtikov V I and Varentsov D 2009 *AIP Conf. Proc.* **1195** 667
[10] Gottschalk B, Koehler A M, Schneider R J, *et al.* 1993 *Nucl. Instr. Meth.* B **74** 467
[11] Agostinelli S, Allison J, Amako K, *et al.* 2003 *Nucl. Instr. Meth.* A **506** 250
[12] Stöcker H and Sturm C 2011 *Nucl. Phys.* A **862-3** 97
[13] Durante M and Cucinotta F A 2011 *Rev. Mod. Phys.* **83** 1245
[14] Solov'yov A V, Surdutovich E, Scifoni E, Mishustin I and Greiner W 2009 *Phys. Rev.* E **79** 011909
[15] Köhler A M 1968 *Science* **160** 303
[16] Han B, Xu X G and Chen G T 2011 *Med. Phys.* **38** 1903
[17] Varentsov D, Bogdanov A, Demidov V S, *et al.* 2012 *Phys. Med.* Apr.1 [epub ahead of print]
[18] Krämer M and Durante M 2010 *Eur. Phys. J.* D **60** 195
[19] Obolensky O I, Surdutovich E, Pshenichnov I, Mishustin I, Solov'yov A V and Greiner W 2008 *Nucl. Instr. Meth.* B **266** 1623

Journal of Physics: Conference Series **373** (2012) 012017 doi:10.1088/1742-6596/373/1/012017

Ion beams in radiotherapy - from tracks to treatment planning

M. Krämer[1], **E. Scifoni**[1], **C. Wälzlein**[1] **and M. Durante**[1]

[1] GSI Helmholtzzentrum für Schwerionenforschung, Planckstr. 1, D-64291 Darmstadt, Germany

E-mail: M.Kraemer@gsi.de

Abstract. Several dozen clinical sites around the world apply beams of fast light ions for radiotherapeutical purposes. Thus there is a vested interest in the various physical and radiobiological processes governing the interaction of ion beams with matter, specifically living systems. We discuss the various modelling steps which lead from basic interactions to the application in actual patient treatment planning. The nano- and microscopic scale is covered by sample calculations with our TRAX code. On the macroscopic scale we feature the TRiP98 treatment planning system, which was clinically used in GSI's radiotherapy pilot project.

1. Introduction

At present about thirty clinical sites around the world apply beams of fast light ions (protons and/or carbon) for the purpose for radiotherapy, and several more are under construction or planned [1]. A major issue of modern radiotherapy is the delivery of a sufficiently high dose to the target volume whereas the exposure of healthy tissue should be kept at a minimum. Fast light ions offer significant physical and radiobiological advantages over conventional photon radiation to achieve this goal, as evidenced for example in GSI's pilot project on radiotherapy with carbon ions [2]. However, application of high-LET radiation requires sophisticated and efficient dose calculation and optimization procedures to obtain acceptable patient treatment plans. The various steps to achieve this goal will be described bottom-up. Micro- or even nanoscopic description of energy, or, better yet, damage deposition provides important insights into the basic nature of radiation action at a very low level. Radiobiological modelling, for example, relies to a great extent on a sufficiently accurate description of dose distributions in the sub-micron range. The complexity of biological systems, however, usually prevents complete mechanistic calculations of treatment plans due to the lack of knowledge about important parameters. Therefore, macroscopic descriptions of dose deposition and biological radiation action are currently the methods of choice for radiotherapeutical applications. Various aspects of dose optimization will be discussed and last but not least means of experimental validation will be presented.

2. Ion tracks - first principles

In view of the stochastic nature of the basic ion-matter interactions it is tempting to use nanoscale calculations based on the Monte Carlo method to describe energy - or better: damage deposition at the lowest possible level. Various such attempts exist, for example the code

PARTRAC [3] which is dedicated to describing DNA damage or the GSI grown TRAX [4, 5], suited for general calculations on ion track structure and many more related quantities. These codes apply a set of cross sections for elementary interactions, typically elastic scattering, ionization, excitation, in contrast to traditional approaches relying on "condensed random walk", as introduced already in the 1960s [6]. Figure 1 for example shows simulations of δ-electrons created by ions of various energies. This representation, however, exhibits only the interaction

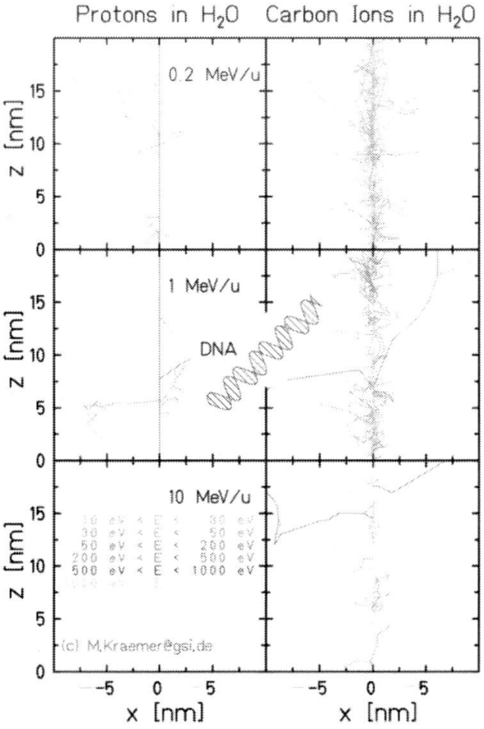

Figure 1. Paths of δ-electrons created by ions of various energies in water, simulated with TRAX. The dense emission pattern of low energy ions can be seen. The path segments are coloured according to electron energy. Most electrons have very low energies too, below 100eV. The inserted DNA schematics shows the scale of the simulation, i.e. nanometers.

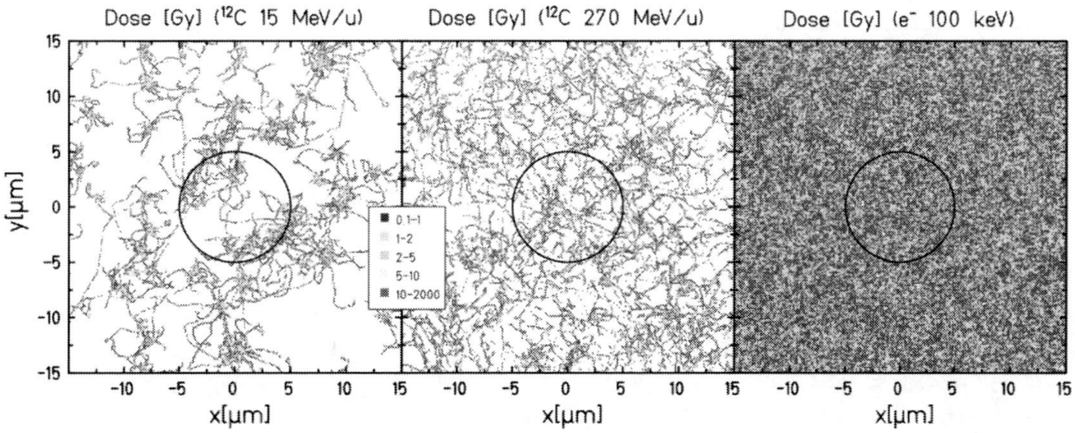

Figure 2. Microscopic dose deposition of low and high-LET radiation. The circle indicates a cell nucleus.

geometry of single tracks in relation to sensitive structures like the DNA. A more realistic scenario is depicted in figure 2, where the small-scale energy deposition pertaining to the same macroscopic dose ($\approx$ 1 Gy) is simulated for radiation of different ionization densities. Low-LET radiation yields an almost homogeneous distribution, whereas stopping particles (^{12}C at 15 MeV/u) show track cores with local doses in excess of 10 Gy. Given that biological systems usually respond linear-quadratically as a function of dose, it does not come as a surprise that such high-LET radiation will have an elevated biological effect and thus will be suited as a means for radiotherapy.

With the ability to describe the physical radiation action on the micro- or even nanoscopic level, it might be tempting to describe also the biological action from the first principles. One of the problems of classical microdosimetry, however, is that there is no obvious link from physical dose distributions to resulting biological damage, and this is also true for most if not all models relying on basic physics alone. Figure 3, for example, exhibits a "show stopper". Survival curves after exposure to various kinds of radiation were measured for two widely used cell lines, one being a repair-deficient mutant of the other, so only a few genes are different. That is, the physics is the same, but the biological outcome is entirely different. Nevertheless several groups attempted a mechanistic modelling approach towards a realistic description of biological damage. For an exemplary but incomprehensive list, we refer to a dedicated topical issue [7] as well as to other contributions in the present issue. For example, the complex chemical stage has been included [8] and after reproducing simple plasmid DNA double strand breaks [9] even repair mechanisms [10] have been considered. However, for reaching the predictive power required in the clinical stage, i.e. predicting arbitrary biological endpoints for entire organs a priori, several more years of experimental and modelling efforts would be necessary.

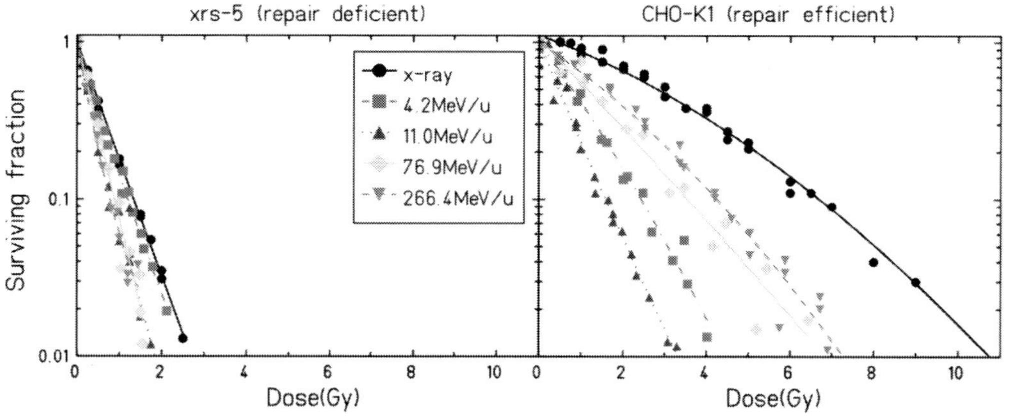

Figure 3. Response (cell survival) of repair-deficient XRS-5 vs repair-efficient CHO-K1, for x-rays and ^{12}C ions with the indicated energies [12].

3. Radiobiological modelling in treatment planning

For all ions heavier than protons (and, strictly speaking, even for protons) the inclusion of radiobiological effects, such as survival probabilities and biologically effective dose, into regular treatment planning is indispensable. Figure 4 illustrates the typical response of a cellular system to different types of radiation. Cell survival, S, usually is described as a linear quadratic dose response, $-\ln S = \alpha D + \beta D^2$, with coefficients α, β depending on cell type, biological endpoint and type of radiation (ion species). The Relative Biological Effectiveness, RBE, is defined as the ratio $RBE = D_x/D_{ion}$ of a reference (i.e. photon) dose, D_x, to an ion dose, D_{ion}, leading to the same effect. As figure 4 illustrates, RBE is not a fixed property of the respective radiation, but rather depends on the chosen survival level and hence on the absorbed dose. The target quantity for treatment planning is the RBE-weighted dose, sometimes also called "biological dose", D_{bio}, connected to the absorbed (or physical) dose via $D_{bio} = RBE(D_{abs}) \times D_{abs}$. As mentioned,

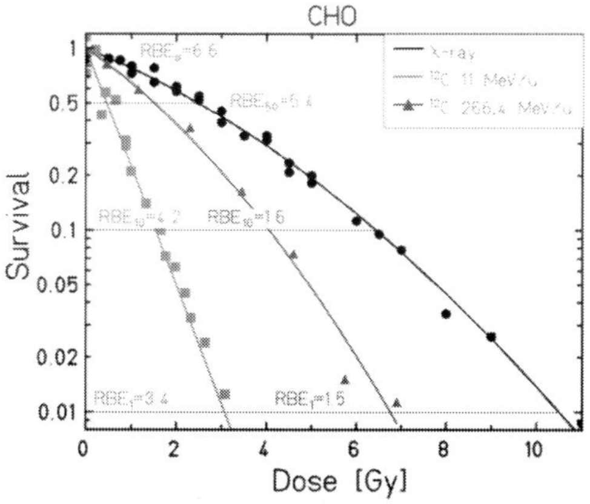

Figure 4. Definition of the RBE on the basis of survival curves for CHO cells exposed to different kinds of radiation. The dependence of RBE on survival and dose level is clearly visible.

a full, bottom-up nanoscopic modelling of radiation damage for the purpose of radiotherapy does not appear to be feasible in the near future. Given the dependence on a multitude of parameters, a pragmatic approach like the one taken in the GSI pilot project is to be preferred. In these kinds of approach, nevertheless, the nanoscopic modelling still plays a relevant role, in describing the physics up to a certain scale, to the highest possible detailed level. In fact, a crucial quantity here is the detailed description of dose deposition around an ion path which is represented by radial dose distributions, a collection of which is shown in figure 5 for various ion energies. The complex biological radiation action can be described by the versatile Local

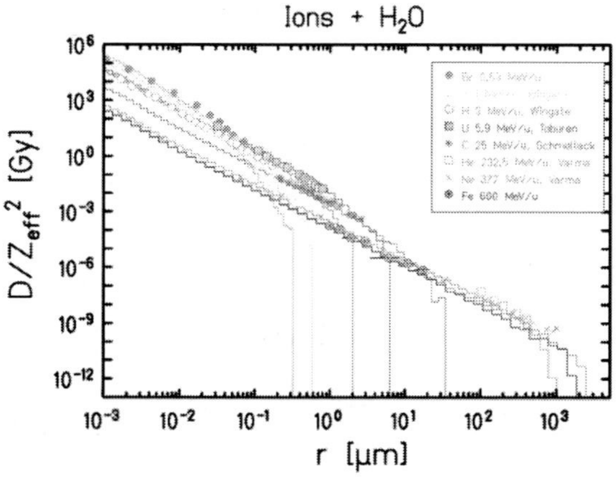

Figure 5. Radial dose distributions, D(r), for ions of various Z and E: TRAX calculations in comparison with the few available experimental data.

Effect model (LEM) [13,14]. It is so far the most successful radiobiological model in clinical use in radiotherapy. Its principles are illustrated in figure 6.

The basic assumptions are that at the local level the radiation damage by sparsely ionizing photon radiation is the same as for ion radiation, and that the cell nucleus is the sensitive target. It should be noted that in its most recent versions, the LEM includes more mechanistic details, such as the spatial distribution of double strand breaks and clustering effects [11].

These assumptions allow us to separate the difficult biological aspect of the problem from the purely physical one. The former is represented by the empirical photon dose response curve, $-\ln S = \alpha_x D + \beta_x D^2$, the latter by the microscopic radial dose distribution, $D(r)$. α_x, β_x

usually can be derived from a rich set of experimental data, not only for cell systems in vitro, but also for complete organs in vivo. Combining these main ingredients by integrating the dose response over the cell nucleus yields the response of a cellular system when exposed to ions of a particular type and energy, $-\ln S = \alpha_z D_z + \beta_z D_z^2$, where D_z is the specific energy deposited in the cell nucleus. It should be emphasized that LEM only provides intrinsic α_z, β_z for single monoenergetic particle traversals in the first place. Computation of biological effects in complex radiation fields, as they are common in radiation therapy, is the task of the treatment planning system (TPS), such as GSI's TRiP98 [5, 15]. This task, i.e. to determine the coefficients α_D, β_D in $-\ln S = \alpha_D D + \beta_D D^2$ in each voxel of the irradiated tissue, is not trivial, but two solutions exist, a "classical" one, described in [16], and a much faster approximation in [17]. Once $-\ln S$ is known, all other quantities, including RBE, can be deduced.

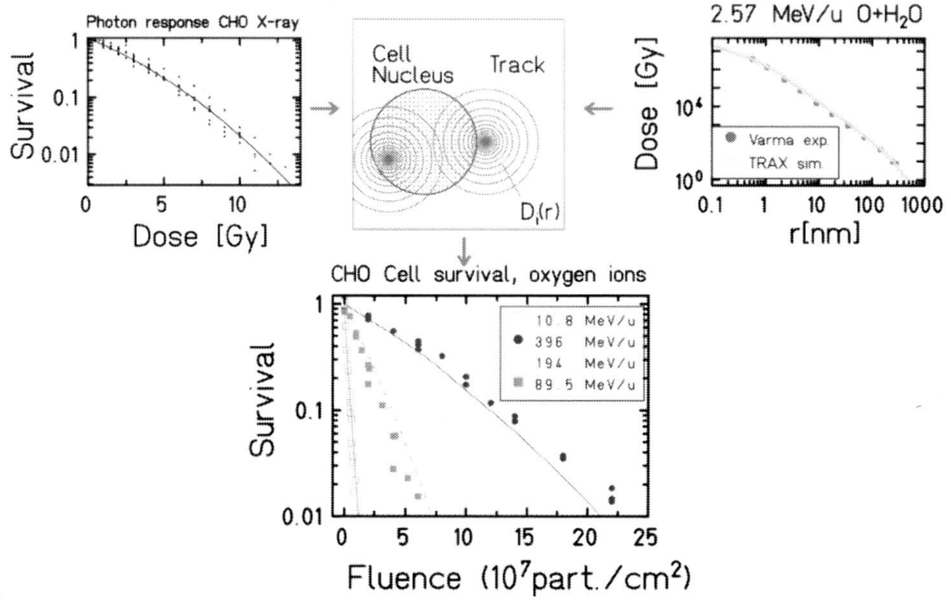

Figure 6. Illustration of the LEM principles: experimental photon dose response combined with radial dose distributions around an ion track on a cell nucleus yields survival curves.

4. Beam application and dose optimization

Apart from the ion accelerator, the centrepiece of a modern ion beam delivery is a magnetic raster scanner (figure 7), pioneered by the GSI pilot project [18]. These devices provide completely active dose delivery by means of lateral magnetic deflection of pencil beams with a selectable FWHM in the range of 4 to 10 mm. Energy variation of the ion beams by the accelerator allows us to cover penetration depths up to 30cm in water without energy degraders, typically in ≈250 steps. A typical single treatment field thus may comprise as many as 50000 individual pencil beam spots organized in ≈ 50 different energy slices. One of the main tasks of treatment planning is to determine the number of particles, $\vec{N}$, within each pencil beam so that the resulting RBE-weighted dose distribution,

$$D_{bio}(\vec{x}, \vec{N}) = D_{abs}(\vec{x}, \vec{N}) \times \mathrm{RBE}(\vec{x}, \vec{N}). \quad (1)$$

matches the medical prescription. Mathematically this can be formulated as a least-squares minimization problem, which reads in simplified form (omitting weight factors):

$$\chi^2 = \sum_{\vec{x}_{target}} (D_P(\vec{x}) - D_A(\vec{x}))^2 + \sum_{\vec{x}_{OAR}} (D_{OAR}(\vec{x}) - D_A(\vec{x}))^2 \times \theta(D_A(\vec{x}) - D_{OAR}(\vec{x})) = min. \quad (2)$$

where $D_P(\vec{x})$ and $D_A(\vec{x})$ are the prescribed and actual dose distribution, respectively, and $D_{OAR}(\vec{x})$ is the maximum allowed dose level in the Organs at Risk (OAR). θ denotes the Heaviside function, its purpose is to impose a penalty if $D_{OAR}(\vec{x})$ is exceeded, but to do nothing if the actual dose is below that limit. The sum runs over all voxels in the target and the OAR, respectively. The solution of equation (2) is not trivial due to the nonlinear dependence on $\vec{N}$ and the constraint term, but can be achieved by appropriate iterative algorithms implemented in the treatment planning system (TRiP98, [19]).

Sample distributions of the resulting particle numbers $\vec{N}$ are shown in figure 8. They are necessarily inhomogeneous to account for the preirradiation of proximal slices by distal ones.

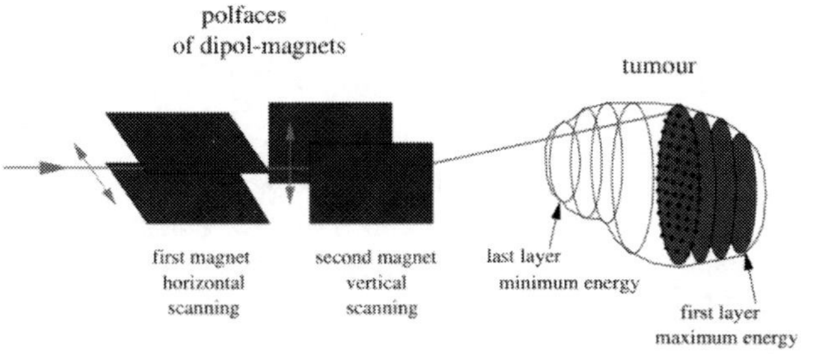

Figure 7. Schematic view of a magnetic raster scan system.

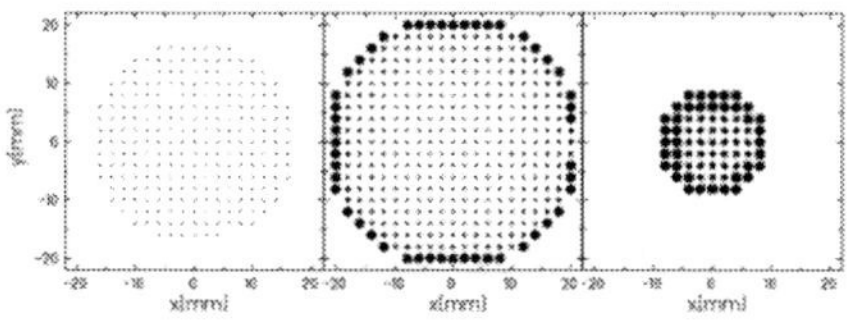

Figure 8. Inhomogeneous particle distribution, $\vec{N}$, for various slices in the target (tumour) volume.

5. Patient plans and dose verification

Figure 9 shows a resulting patient plan. The high selectivity of carbon ions together with the sophisticated dose optimization procedure allows to cover the tumour volume with the prescribed dose and to spare the OAR to a high degree.

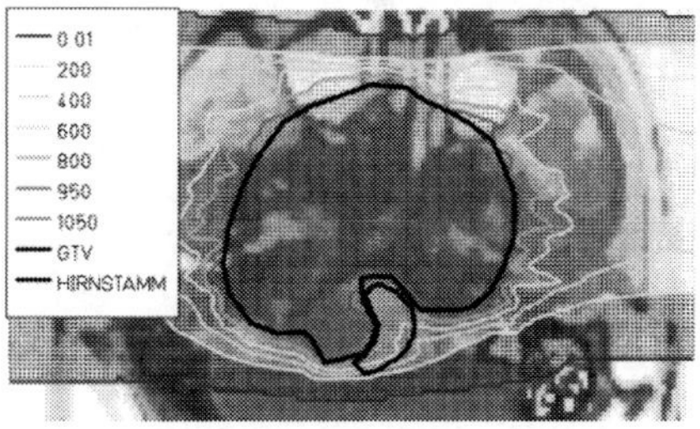

Figure 9. RBE-weighted (biological) dose distribution covering a large tumour (chordoma) surrounding an organ at risk (the brain stem) using two simultaneously optimized opposing fields. The colours and numbers refer to relative dose levels in per mil of planned dose, i.e. 1000 corresponds to an RBE-weighted dose of 3 Gy.

The validity of actual patient plans is usually verified by measuring the absorbed dose distributions in a water phantom [20]. A plan will be accepted if measured and TPS-predicted dose distributions differ by no more than a certain amount, five percent being a typical value.

The ultimate goal of radiotherapy, however, is to achieve a biological effect rather than just delivering an absorbed dose distribution. Thus, it is a good idea to verify the TPS calculations by means of biological dosimetry, i.e. by measuring cell survival distributions after irradiation. Since

Journal of Physics: Conference Series **373** (2012) 012017 doi:10.1088/1742-6596/373/1/012017

this is not possible for each patient in vivo, a so-called Bio-Phantom (figure 10) was developed at GSI which is used to verify typical treatment situations under patient-like conditions.

Figure 10. The biophantom device used to measure cell survival distributions. It consists of a vessel with $\approx$ 30cm diameter, filled with medium. Cells are deployed on a set of plastic slabs submerged in the medium at well-defined positions, thus serving as a two-dimensional detection system. The whole setup is positioned on the couch and irradiated the same way as a patient.

Figure 11 shows a biodosimetry example for a 2-field optimized plan, a C-shaped target volume wrapping around an assumed OAR. Although CHO cells are not direct surrogates for human tissue, both biological systems follow the same rules as far as radiation damage is concerned, and thus CHO can serve well for the purpose of biological verification of complex irradiation plans.

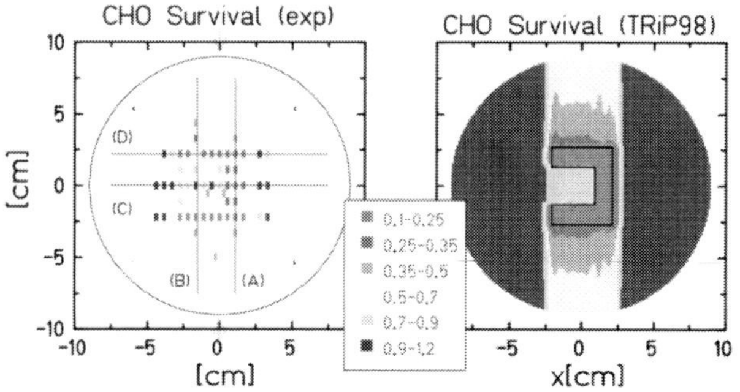

Figure 11. Top view of two-dimensional survival distributions of Chinese hamster ovary (CHO) cells, measured with the Bio-Phantom (left) and calculated with the TRiP98 TPS (right). Comparison by colour shows the agreement between measurement and calculation. The numbers associated with the different colours denote the surviving fraction.

6. Conclusion and Outlook

We have described the various steps which lead from a microscopic picture of radiation action to "real life" applications, i.e. patient treatment with ion beams. The described models and methods were successfully applied in the treatment of more than 430 patients from 1997 to 2008 in the GSI pilot project. Follow-up projects like HIT and CNAO are now applying these methods on a larger clinical scale.

TRiP98 and following releases of the TPS are well suited to serve as research prototypes for ion-beam radiotherapy planning, in particular for 4D treatment [21], as well as for so-called "Adaptive Treatment Planning", aimed at the improvement of dose optimization in case of rapid spatio-temporal variation of organ geometry or tumour radiosensitivity. As an example for ongoing research, figure 12 shows model calculations on the so-called Oxygen Enhancement Ratio (OER), which is a measure for the radio-insensitivity of tissue with reduced oxygen content. The TPS is currently enhanced to provide compensation for this "oxygen effect" [22].

At the other end of the scale, the TRAX simulation code will be enhanced towards lower distances, since the track core plays an import role in radiobiological modelling. Figure 13 shows radial dose distributions followed to lower distances in various scenarios [23]. Due to a

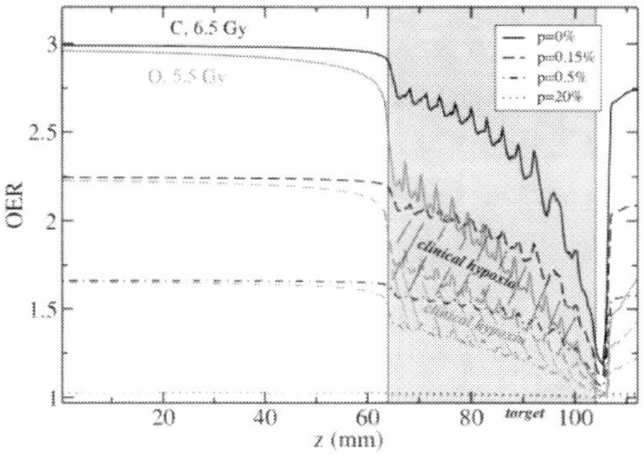

Figure 12. Oxygen enhancement ratio as a function of depth (with relative oxygen content as a parameter) for carbon and oxygen ions delivering the same dose to the entrance tissue. The shaded area indicates a potential target volume. For very high-LET radiation (i.e. stopping oxygen ions) the OER approaches unity, i.e. the oxygen effect vanishes.

lack of experimental data in condensed phase, however, an independent verification currently is difficult.

Lower distances are inevitably related to lower $\delta-$electron energies. Thus we also strive towards better low-energy cross sections and towards target materials other than water, in context with the efforts to measure low-energy electron emission from condensed matter with the GSI Toroid spectrometer [24].

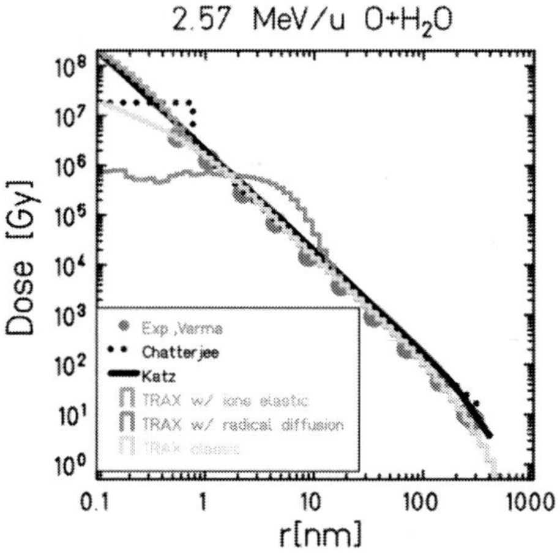

Figure 13. Experimental radial dose distribution (in gaseous medium) compared with TRAX simulations (in water) under various assumptions. Accounting for the small-angle elastic scattering of the primary ions results only in small differences to the classic calculations. In contrast, inclusion of radical diffusion, which is present in living systems, significantly affects the radial dose distribution.

References

[1] *http://ptcog.web.psi.ch/ptcentres.html*
[2] Schardt D, Elsässer Th and Schulz-Ertner D 2010 *Rev. Mod. Phys.* **82** 383
[3] Friedland W *et al* 2008 *Radiat. Environ. Biophys.* **47** 49
[4] Krämer M and Kraft G 1994 *Radiat. Environ. Biophys.* **33** 91
[5] Krämer M, Durante M 2010 *Eur. Phys. J.* D **60 (1)** 195
[6] Berger MJ 1963 *Methods Comp Phys* **1** 135
[7] Baccarelli I *et al* 2010 Topical issue on Molecular Level Assessments of Radiation Biodamage *Eur. Phys. J.* D **60** 1
[8] Nikjoo H *et al* 2006 *Rad Meas* **41** 1052
[9] Grosswendt B *et al* 2005 *Radiat. Prot. Dosimetry* **115** 1
[10] Friedland W *et al* 2012 *Int. J. Radiat. Biol.* **88** 129
[11] Friedrich T *et al* 2012 *Int. J. Radiat. Biol.* **88** 103
[12] Weyrather WK, Ritter S, Scholz M and Kraft G 1999 *Int. J. Radiat. Biol.* **75(11)** 1357
[13] Scholz M, Kellerer AM, Kraft-Weyrather W and Kraft G 1997 *Radiat. Environ. Biophys.* **36** 59
[14] Elsässer Th *et al* 2010 *Int. J. Radiat. Oncol. Biol. Phys.* **78 (4)**
[15] Krämer M, Jäkel O, Haberer T, Kraft G, Schardt D and Weber U 2000 *Phys. Med. Biol.* **45** 3299

[16] Krämer M, Scholz M 2000 *Phys. Med. Biol.* **45** 3319
[17] Krämer M, Scholz M 2006 *Phys. Med. Biol.* **51** 1959
[18] Haberer T *et al* 1993 *NIM A* **330** 296
[19] Gemmel A, Hasch B, Ellerbrock M, Weyrather WK and Krämer M 2008 *Phys. Med. Biol.* **53** 6991
[20] Karger CP, Jäkel O, Hartmann G H Heeg P 1999 *Med. Phys.* **26(10)** 2125
[21] Bert Ch and Durante M 2011 *Phys. Med. Biol.* **56** R113
[22] Scifoni E, Krämer M and Durante M 2011 *http://www.gsi.de/library/GSI-Report-2012-1*
[23] Wälzlein C, Krämer M, Scifoni E and Durante M 2011 *http://www.gsi.de/library/GSI-Report-2012-1*
[24] Lineva N, Kozhuharov C, Hagmann S, Krämer M and Kraft G 2009 *Nucl. Instr. and Meth.* B **267** 891

Hadrontherapy - macrobenefit in cancer therapy?

J L Habrand, E Baron, J Bourhis, J Datchary, A Mazal and K Meflah[1]

ARCHADE Project for Hadrontherapy, Caen, France

E-mail: jl.habrand@baclesse.fr

Abstract. Hadrontherapy is one of the most promising radiotherapeutical innovations that deal with accelerated heavy charged particles, mainly proton and carbon ions. Their salient features include an original dose-distribution, based on the Bragg curve, and in some of them an increased RBE at the range-end. Approximately 100 000 patients have been treated so far in approximately 40 centers worldwide. Outstanding outcomes have been substantiated in rare neoplasms using protons, such as ocular melanomas, skull base sarcomas, and pediatric malignancies, while only promising evidences have emerged using carbons. Assessing their place in more common tumor-sites, such as lung, pancreas, prostate, esophagus remains to be determined, and justifies the expansion of future particle therapy programs.

1. Background

Radiation therapy (RT) along with surgery represents the mainstay of local treatment of cancer. In France, approx 180 000 patients are treated annually using RT, and 600 000 in Europe. On the other hand, it is established that an efficient local treatment can, to some extent, prevent a metastatic evolution of the disease. In this context, RT has enjoyed considerable improvements since the early 90s that make it safer and more efficient. Most of these innovations have become possible because of dramatic developments of imaging (especially 3D imaging) that considerably benefit treatment preparation and verification, and computer technologies that are part of all modern equipment (including linear accelerators, gantries, robotised couches...). Public attention has been recently attracted by the capabilities for producing a modulated intensity of the beam (IMXRT), a sophisticated delivery mode that allows strict conformation of multiple beams, "shaped" dynamically in modulated fluence to the target using a "reverse" treatment planning (algorithms take into account specified dose and volume constraints for tumor, and normal organs to generate beams' arrangement and intensity). Helical tomotherapy and cyberknife, the most advanced technologies in this field, are also derived from IMXRT [1]. Nonetheless, all these technologies are based on a single type of radiation: photons (i.e. X-Rays) that have been in use for over a century! Limitations of photon beams are correlated with their physical properties in matter characterised by a progressive attenuation, a substantial "exit" dose at the end of path, and finally unnecessary exposure of anatomical structures beyond the target. Furthermore, part of the tumors are known to be poorly sensitive to photons, (due to cell intrinsic radioresistance, fast multiplication, hypo-oxygenation…), while most normal tissues express immediate and/or delayed radio-induced toxicity, above a specific dose-level. These limitations have stimulated the interest for new approaches that improve tumor sensitivity (through combinations with cytotoxic drugs, biological agents…), or normal tissues tolerance (by radioprotectant agents,

[1] To whom any correspondence should be addressed.

hyperfractionation of the dose…), or both: in this setting, the introduction of hadrontherapy (based on accelerated "heavy charged particles") represents a breakthrough due to outstanding ballistical properties, combined with biological advantages in some of them.

2. A brief history of the charged particle therapy world

In 1946, R Wilson, a physicist at the Harvard Cyclotron (MA, USA) wrote a seminal paper in which he pointed out that a high energy proton beam could be effective for cancer therapy since it produced a greater ionisation density at the end of its path. This physical property has actually been described earlier on (1904) by WH Bragg, and is since known as the Bragg curve or simply the "Bragg peak" (BP). This paper triggered early biological and later clinical experiments, first conducted by a few neurosurgical teams. Remarkably, these experiments represented a limited part of large basic physics programs. In the seventies it became evident that most conventional cyclotrons and synchrocyclotrons (SC) were obsolete for nuclear physics research. But in their energy-range (approx 150 MeV: 16 cm penetration in water) particles could reach many deep-seated tumor sites. Consequently, ophthalmological and neuro-oncological clinical programs quickly developed under the supervision of physicians and physicists initially at Massachusetts General Hospital (MA, USA). Similar developments followed in foreign countries, including France where in 1991, the nuclear physicists "passed the baton" to radiation oncologists at the 201 MeV Orsay SC. In parallel, more limited programs dealing with light ions (neon, carbon, silicium…) were conducted at BEVALAC (CA, USA) until the unfortunate shut-down of all physics and medical applications in 1993. We will also mention the negative pions which were piloted in two centers (TRIUMPH, Vancouver, Canada, and PSI, Villigen, Switzerland). Eventually, the message was passed that carbon ions could represent the optimal particle in terms of ballistical and biological advantages. This stimulated intensive investigations dealing with carbon both in Japan (HIMAC, Chiba), and Germany (GSI, Darmstadt). In the 2000s sustained interest in protontherapy applied to new indications led to the development of modern turn-key facilities, developed by a few industrial companies located in Japan, Belgium, and Germany. Although synchrotron technology gained wide acceptance, a renowned interest for innovative compact high energy cyclotrons emerged recently. Similarly, synchrotron-based development of carbon centers was achieved on a more modest scale in a few places (located in Japan and Germany, so far), although the cyclotron technology has its own supporters. It is also remarkable that recent modes of dose-delivery have been mimicking advanced photon innovations (i.e. isocentric gantries, multi leaf-collimators, spot-scanning applied to intensity modulation…) so that hadron-patients can benefit from the highest levels of sophistication.

3. Clinical benefit of hadrontherapy

Following pioneering work initiated in the 60s in the US, two radiation types have been tested since the early 90s: protons and "light" ions, mainly carbon ions. As off December 2011, 95 000 patients (most with cancers) had benefited from hadrontherapy in 37 centers worldwide, including 31 proton, 4 carbon, and 2 mixed proton and carbon centers[3].

- **Protons** are well adapted to high precision irradiation, due to their Bragg curve dose distribution: the selective absorption of the dose at tumor depth is followed by a sharp fall-off. The lateral "penumbra" is reduced as well compared with most photon dose profiles [4]. All these properties improve sparing of normal tissues located downstream or lateral to the beam (even away by few millimetres only), and allow dose-escalation to the target more safely if improved tumor cell-killing is also contemplated [5]. To some extent, this type of radiation is the best that complies with the ALARA (As Low As Reasonably Achievable) principle that has been recommended for the general population exposed to radiation hazards, esp. in terms of radio-associated carcinogenesis.

- *Clinical studies* have mainly concerned tumors poorly sensitive to radiations, located close to critical sensitive organs, and conducted in approximately 85 000 patients. They are summarized in table 1. Ocular melanomas represent such a paradigm due to their well-known resistance to X-Rays and vicinity of critical anatomical structures (macula, optic disk...). On approximately 10 000 patients studied, a local control in excess of 95% and a survival rate of 80% have been reported, along with good cosmetic and visual preservation [6]. Slow-growing primary malignancies of skull base and spine (esp. chordomas and chondrosarcomas), although rare conditions, represent a true therapeutical challenge due to their radioresistance, incomplete surgical resectability, and frequent encasement in vital structures (brain stem, spinal cord...). Local control using protons combined with partial resection, has significantly improved the outcome and almost doubled it in poor prognostic subgroups (from approx. 35 to 70%) [7,8]. Pediatric tumors, another rare morbid condition, are also particularly challenging due to the exquisite sensitivity to ionizing radiation of growing normal organs, and documented cosmetic, functional, and sometimes lethal, side-effects observed years or decades following RT. Although chemotherapy still retains the major role in the therapeutical armamentarium, proton therapy (PT) seems optimal, when RT is still indicated [9]. Recently, more common tumor-types such as early lung, or prostate carcinomas have been widely tested, esp. in the US and Japan [10-13]. Encouraging results have been reported, including positive randomized trials supporting the benefit of and good tolerance to high dose PT [14]. An interesting finding was also the improved tolerance to concomitant chemo-radiation combinations, an efficient but highly toxic approach, commonly administered in oncology [14]. Pre-clinical dosimetrical investigations are also paving new avenues to emerging indications, so that the estimated number of patients suitable for PT (sorted out recently by a foreign group) could approach 10 to 15% of all RT patients [15]. This would correspond to 20 000 yearly in France, including 3 500 in the Western and Northern part of the country. This number would by far exceed the current capacities of the two existing French treatment centers (30 or so, worldwide).

Table 1: Relevant clinical series on proton therapy. *Abbreviations: dpf: dose per fraction; CGE: Cobalt-Gy equivalent; LC: local control; OS: overall survival.*

Author, date [ref]	*Tumor type, #cases*	*Dose/ dpf (CGE)*	*Tumor control*	*Toxicity*
Dendale, 2006 [6]	Ocular melanoma *1406*	60/15	LC:96% OS:79%	Enucleation: 7%
Munzenrider, 1999 [7]	Skull base sarcomas *519*	67/1.8	LC:73-98% OS: 80-91%	
Habrand, 2009 [9]	Pediatrics *108*	50-70/1.8	LC: 81% OS: 88%	Severe: 3%
Zietman, 2005 [10]	Early prostate *393*	® 70-79/	LC: 61-80% OS: 96-97%	Severe : 1-2%
Bush, 2004 [11]	Lung *68*	51/5	LC/74% OS/ 44%	
Sugahara, 2005 [12]	Esophagus *46*	76-82/	LC: 57% OS: 34%	
Chiba, 2005 [13]	Liver *162*	72/4.5	LC: 87% OS: 23%	Severe: 3%

Journal of Physics: Conference Series **373** (2012) 012018 doi:10.1088/1742-6596/373/1/012018

- **Light ions:** as mentioned above, initial biological and clinical experiments conducted in North America, led to the choice of carbon ions ($_{12}C$) as the optimal ion species in this field. Further experiments switched to different geographical places (Germany and Japan), including a recent German high tech center (HIT, Heidelberg), and five additional ones in Japan. These high LET particles share, with protons, a high ballistical selectivity, associated with the BP, and indicate additional remarkable biological properties, correlated with increased energy-deposition and nuclear interactions at end of range. Quantitatively, these correspond to an increased Relative Biological Efficiency (RBE) from approx. 1.1 in the entrance channel (where normal tissues are interposed), up to 3 and more in the spread-out BP (where tumor and BP superimpose).

Clinical experience so far, dealing with carbon ions, has been reported for approximately 9 000 patients worldwide, and 1 500 using other ion species (mainly helium and pions) in initial experiments. Carbon ions have been chiefly delivered with peculiar fractionation regimens of the dose, i.e. hypofractionated. Clinical indications are summarized on table 2. They concern mainly radio-resistant advanced malignancies, located close to critical structures. If the benefit over PT remains unclear in its main indications, additional interesting data have been collected in highly radio-resistant tumor-processes, such as mucosal melanomas, bony and soft tissue sarcomas, salivary gland carcinomas, digestive tract primaries...[16-22]. Noteworthily appears the impact on pancreatic carcinoma, an almost uniformly lethal condition. A recent update of the Japanese experience, showed a 5 year survival approaching 100% in selected sub-groups [19]. Re-irradiation of a locally recurrent tumor, following previous irradiation, is another challenging situation (due to poor tolerance of tissues irradiated twice, and poor sensitivity related with hypoxia) in which carbon ions have been tested with promising results. A proton course followed by a carbon boost has also been suggested in most radioresistant processes located in critical anatomical areas. Randomised trials comparing carbon ions

Table 2: relevant clinical series on carbon therapy. *Abbrev: see table 1*

Author, date [ref]	*Tumor-type, #cases*	*Dose/dpf*	*Tumor control*	*Toxicity*
Kamada, 2002 [16]	Advanced bone+soft part sarcomas *64*	53-74/3.3-4.5	LC: 88% OS: 46%	Severe: 9%
Schulz-Ertner, 2005 [17]	Advanced salivary *29*	72/3	LC: 77% OS: 76%	Severe: 3%
Combs, 2010 [18]	Aggressive meningiomas *10*	68/3	LC: 86% OS; 75%	None
Shinoto, 2011 [19]	Post op pancreas *26*	30-37/4-4.5	LC: 100% OS: 53%	Severe: 8%
Ishikawa, 2006 [20]	Advanced prostate *142*	66//3.3	LC: 99% OS: 88%	No severe
Yamada, 2009 [21]	Recurrent rectal *112*	67-74/	LC: 97% OS: 40%	No severe
Ohno, 2011 [22]	Advanced uterine cervix *22*	64-72/3-9	LC: 68% OS: 50%	No severe if bowel<60 Gy
Ohno, 2011 [22]	Advanced uterine corpus *45*	62-71/3-9	LC: 68% OS: 50%	No severe if bowel<60 Gy

against reference photon arms are also underway [23-25]. Two ion-projects have been elaborated in France (ETOILE in Lyon, and ARCHADE in Caen) by multidisciplinary teams of physicists, physicians, and engineers in order to design and build-up comprehensive hadron programs "from bench to bedside".

4. Socioeconomic impact

Socioeconomic issues are a primary concern in the context of limited health resources in general and potential cheaper therapeutical strategies in the specific area of cancer. Clearly, hadron program investments are five to ten times higher than their photon counterparts', even those dealing with most advanced technologies. But it was shown that a proton project amortization could be safely achieved if patient accrual were sufficient, by approximately doubling the cost per session of photons [26]. As treatment cost is conditioned at least in part by the total number of sessions, light ions provide additional cost-savings by reducing it: current protocols dealing with carbon in Japan and Germany administer generally the dose in 1 to 3 daily sessions per week (vs 5 for protons and photons). The total number of fractions, usually 30 and more in "conventional" fractionation, is also dramatically reduced in most studies, especially in the Japanese approach (to a total of 4 to 16). In the long run, several aspects also speak in favor of particles and make them attractive: costs of salvage programs and patients' rehabilitation can be minimized; also prolonged employment together with a good quality of life is expected [27]. Detailed evaluations have been implemented in this field.

References

[1] Levitt SH, Purdy JA, Perez CA and Poortmans P 2012 *Technical basis of radiation therapy. Practical clinical applications*. 5th Edition. Ed. Springer-Verlag Pub. Berlin
[2] Welsh J, Gomez D, Palmer MB, Riley BA, Mayankkumar AV, Komaki R, Dong L, Zhu XR, Likhacheva A, Liao Z, Hofstetter WL, Ajani JA and Cox JD 2011 *Int J Radiat Oncol Biol Phys.* **81** 1336
[3] http://ptcog.web.psi.ch/patient_statistics.html
[4] Mazal A, Habrand JL, Delacroix S, Datchary J, Dendale R, Desjardins L, Ferrand R, Malka V and Fourquet 2010 *A Bull Cancer* **97** 831
[5] Bouyon-Monteau A, Habrand JL, Datchary J, Alapetite C, Bolle S, Dendale R, Feuvret L, Helfre S, Calugaru V, Cosset JM, and Bey P 2010 *Cancer Radiother.* **14** 727
[6] Dendale R, Lumbroso-Le Rouic L, Noel G, Feuvret L, Levy C, Delacroix S, Meyer A, Nauraye C, Mazal A, Mammar H, Garcia P, D'Hermies F, Frau E, Plancher C, Asselain B, Schlienger P, Mazeron JJ and Desjardins L. 2006 *Int J Radiat Oncol Biol Phys.* **65** 780
[7] Munzenrider JE and Liebsch JN. 1999 *Strahlenther Onkol.* **175** 57
[8] Noël G, Feuvret L, Calugaru V, Dhermain F, Mammar H, Haie-Méder C, Ponvert D, Hasboun D, Ferrand R, Nauraye C, Boisserie G, Beaudré A, Gaboriaud G, Mazal A, Habrand JL and Mazeron JJ 2005 *Acta Oncol.* **44** 700
[9] Habrand JL, Bolle S, Datchary J, Alapetite C, Petras S, Helfre S, Feuvret L, Calugaru V, De Marzi L, Bouyon-Monteau A, Dendale R, Kalifa C, Grill J and Doz F. 2009 *Cancer Radiother.* **13** 550
[10] Zietman AL, DeSilvio ML, Slater JD, Rossi CJ Jr, Miller DW, Adams JA and Shipley WU 2005 *JAMA.* **294** 1233
[11] Bush AD, Slater JD, Shin BB, Cheek G, Miller DW and Slater JM 1999 *Chest* **116** 1313
[12] Sugahara S, Tokuuye K, Okumura T, Nakahara K, Saida Y, Kagel K, Ohara K, Hata M, Igaki H and Akine Y. 2005 *Int J Radiat Oncol Biol Phys* **61** 76
[13] Chiba T, Tokuuye K, Matsuzaki Y, Sugahara S, Chuganji Y, Kagei K, Shoda J, Hata M, Abei M, Igaki H, Tanaka N and Akine Y. 2005 *Clin Cancer Res* **11** 3799
[14] Chang JY, Komaki R, Lu C, Wen HY, Allen PK, Tsao A, Gillin M, Mohan R and Cox JD 2012 *Cancer* [Epub ahead of print]

[15] Glimelius B, Ask A, Bjelkengren G, Björk-Eriksson T, Blomquist E, Johansson B, Karlsson M and Zackrisson B 2005 *Acta Oncol.* **44** 836
[16] Kamada T, Tsujii H, Tsuji H, Yanagi T, Mizoe JE, Miyamoto T, Kato H, Yamada S, Morita S, Yoshikawa K, Kandatsu S and Tateishi A. 2002 *J Clin Oncol* **20** 4466
[17] Schultz-Ertner D, Nikoghosian A and Didinger B 2005 *Cancer* **104** 338
[18] Combs SE, Hartmann A, Nikoghosian A, Jäkel O, Karger CP, Haberer T, von Deimling A, Münter MW, Huber PE, Debus J and Schultz-Ertner D. 2010 *Radiother Oncol* **95** 54
[19] Carbon ion radiotherapy for pancreatic cancer. Shinoto M, Yamada S, Imada H, Kamada T and Tsujii H. Proceedings 2nd Joint Symposium 2011 on Carbon Ion Radiotherapy. Abstr, 60-65. Kamada T, Balosso J Ed. NIRS/GCS ETOILE Pub
[20] Ishikawa H, Tsuji H, Kamada T, Yanagi T, Mizoe JE, Kanai T, Morita S, Wakatsuki M, Shimazaki J and Tsujii H. 2006 *Radiother Oncol* **81** 57
[21] Yamada S, Shinoto M, Shigeo Y, Imada H, Kato H, Kamada T and Tsujii H. 2009 *Gan To Kagaku Ryoho* **36** 1263
[22] Ohno T and Kato S 2011 Ion Beam Therapy 2nd Edition U Linz Ed Springer-Verlag Pub Heidelberg
[23] Combs SE, Kalbe A, Nikoghosyan A, Ackermann B, Jäkel O, Haberer T and Debus J. 2011 *Radiother Oncol.* **98** 63
[24] Nikoghosyan AV, Karapanagiotou-Schenkel I, Münter MW, Jensen AD, Combs SE and Debus J. 2010 *BMC Cancer.* **10** 607
[25] Nikoghosyan AV, Rauch G, Münter MW, Jensen AD, Combs SE, Kieser M and Debus J. 2010 *BMC Cancer* **10** 606
[26] Goitein M and Jermann M. 2003 *Clin Oncol.* **15** 37
[27] Lundkvist J, Ekman M, Ericsson SR, Jönsson B and Glimelius B 2005 *Acta Oncol.*, **44** 850.

Development of a low-energy particle irradiation facility for the study of the biological effectiveness of the ion track end

L Manti[1], L Campajola[1], FM Perozziello[1], JN Kavanagh[2], G Schettino[2]
[1]Dipartimento di Scienze Fisiche, Universita' di Napoli "Federico II", Complesso Universitario Monte S. Angelo, 80126, Napoli, Italy
[2]CCRCB, Queen's University Belfast, 97 Lisburn Road, BT9 7BL, Belfast, Northern Ireland, UK

E-mail: g.schettino@qub.ac.uk

Abstract. Uncertainties surround the radiobiological consequences of exposure to charged particles, despite the increasing use of accelerated ion beams for cancer treatment (hadrontherapy). In particular, little is known about the long-term effects on normal tissue at the beam entrance or in the distal part of the Spread-Out Bragg Peak (SOBP). Moreover, although the relative biological effectiveness (RBE) of particle radiation has been traditionally related to the radiation linear energy transfer (LET), it has become increasingly evident that radiation-induced cell death, as well as long term radiation effects, is not adequately described by this parameter. Hence, exploring the effectiveness of various ion beams at or around the Bragg peak of monoenergetic ion beams can prove useful to gain insights into the role played by parameters other than the particle LET in determining the outcome of particle radiation exposures. In this context, the upgrade of the Tandem irradiation facility at Naples University here described, has allowed us to perform a series of preliminary radiobiological measurements using proton and carbon ion beams. The facility is currently used to irradiate normal and cancer cell lines with ion beams such as oxygen and fluorine.

1. Introduction

Ion beams have been used for radiotherapy applications for some years [1], [2], [3] with protons used in many facilities worldwide while heavier ions (i.e. carbon) have been exploited in Japan and Germany with impressive outcomes [4]. Despite these encouraging results, there are still considerable uncertainties regarding the biological effects caused by exposure to charged particles. A deeper understanding of the effects caused by heavy ions on biological samples is therefore essential to develop a rigorous theory of ion radiation action at the cellular and molecular level and improving tumour hadrontherapy.

It is now commonly accepted that the biological effectiveness of charged particles is critically determined by the pattern of ionizations produced by the ion traversal (i.e. track structure) [5]. While the linear energy transfer (LET) is a good indicator of the rate of energy deposition, at extremely high and inhomogeneous LET regions it does not predict biological effects with a simple relationship [6-8]. Radiotherapy studies investigating the biological response of different cell lines to different types of charged particles have focused mainly on the cell killing effect on tumour cells in the so-called Spread Out Bragg Peak (SOBP) [9-11]. However, effects caused in tissues at the beam entrance channel, in the distal part of the SOBP and in the surrounding areas also need to be addressed. Moreover, the increasing interest in utilizing a range of ions, including light ions other than protons, such as boron or lithium, and ions heavier than carbon in therapy [12], demands further development of suitable models (such as the Local Effect Model) to measure and predict the effectiveness of these ions in biological systems.

In this manuscript, we report the upgrading of the radiobiological irradiation facility at the University of Naples to include ions other than protons. Details of the experimental set-up including ion source, dose and energy characterization are reported together with first biological measurements of cell survival, bystander effects and radiation-induced premature cellular senescence in normal human fibroblasts (AG01522) exposed to a proton ion beam. Survival curves following carbon ion irradiation of a panel of glioma human cell lines of varying radiosensitivity are also reported.

2. Materials and Methods

2.1. Irradiation set-up

The radiobiological irradiation facility at the Department of Physical Sciences of the University of Naples "Federico II" is centered on a 3-MV HVEC TTT-3 Tandem accelerator. TiH_2 and graphite cathodes and a negative ion sputtering source are used to produce hydrogen and carbon beams, respectively, which can then be accelerated up to 6 MeV for protons and 21 MeV for carbon. Lithium (Li) and Fluorine (F) ion beams have also been produced using LiF cathodes and accelerated up to 12 MeV and 21 MeV, respectively. Finally, oxygen beams were produced by spraying O_2 gas onto a tantalum cathode; they can be accelerated up to 21 MeV. Due to energy loss in the scattering device and during extraction from vacuum, the maximum ion energy incident onto the cells is slightly lower as indicated in table 1. The beam intensity is monitored by several Faraday cups placed along the beam line and a fast computer controlled beam shutter is positioned after the last cup in order to precisely control the dose delivered to the biological samples. Downstream, a 1 mm collimator is used to focus the beam on the scattering chamber where an Au foil (0.22 μm) scatters the beam achieving the required homogeneous irradiation at the end of the beam line (approximately 2 m away from the scattering chamber, see figure 1). Proton irradiation was performed using the described set-up, in which the beam passed through a 6 μm Mylar window, a 2 mm air gap and the Mylar base (2.5 μm) on which cells were grown. In order to minimize energy loss and thus make it possible to use also heavier ion beams for radio-biologically relevant experiments, the Mylar substrate on which the biological samples are seeded was reduced to 1.5 μm and also used as vacuum window at the end of the beam line. In this way, the air gap and relative uncertainty on energy loss is abolished. This requires a differential vacuum section at the end of the beam line, which can be pumped independently from the main beam line allowing for positioning and removal of the biological samples (figure 1). To prevent mechanical stress on Mylar foil, this is laid upon a tungsten grid of 50 μm wires, which translated in 6% opacity on the irradiated surface.

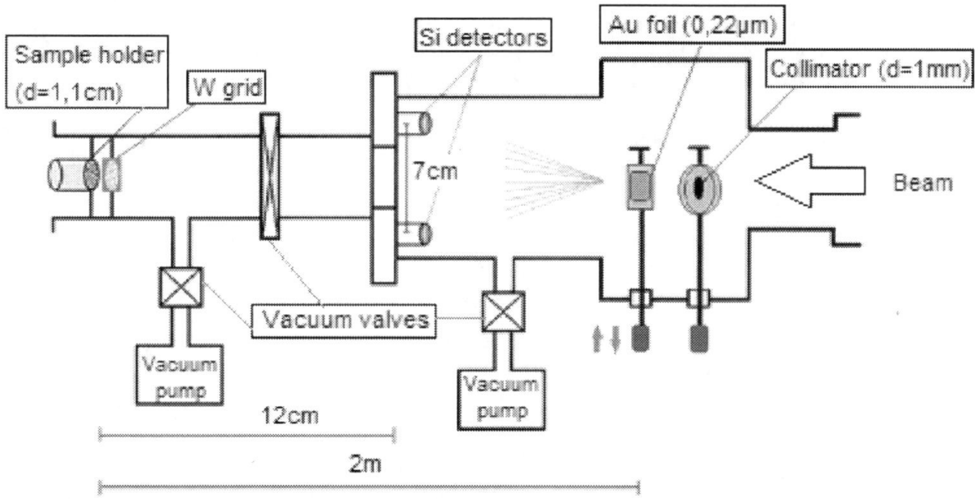

Figure 1. Schematic diagram (not to scale) of the beam line end with differential vacuum sections and the biological sample holder.

2.2. Dosimetry and dose monitoring

Energy spectra and dosimetric measurements are performed 2 m away from the scattering chamber by using Si-based detectors as well as CR39 polycarbonate solid state track detectors. Three Si-based detectors (300 µm thick and 25 mm^2 sensitive area, 14 keV resolution) are amounted at the end of the beam line. The first one is centred with the beam axis and is therefore positioned in the same place where the biological samples will be during the irradiation. This detector is used to characterise the irradiation beam (particle type, energy, fluency) and it is removed during the sample irradiation. The two other Si-based detectors are placed symmetrically at 3.5 cm from the beam axis and are permanently mounted at the end of the beam line for continuously monitoring the dose rate during irradiation. All three Si-based detectors are cross-calibrated in order to estimate beam uniformity and dose delivered to the biological samples. CR39 solid-state track detectors (SSTDs) positioned at the irradiation spot using the biological sample holders (see below) are also used to verify and cross-calibrate the Si-based detector response. Pre- and post-irradiation CR39 counts performed on at least 10 random fields at light microscope (32X, corresponding to a field area of 1.25 10^{-4} cm^2) were used to evaluate the Si-based detector calibration factor and eventually correct retrospectively nominal dose values.

The dose absorbed by the biological sample is calculated using the following formula:

$$\text{Dose (Gy)} = \text{Fluence (particles/cm}^2) \times \text{LET (keV/}\mu\text{m)} \times 1.6\ 10^{-9}.$$

Particle fluence and beam uniformity were measured and monitored by means of the CR-39 SSTDs. From it, administered doses were retrospectively calculated using the calibration factor between Si detector counts and measured mean particle numbers per cm^2 from the CR39 plastics. Etching was carried out by KOH 10N at 80 °C for 30 min. Calculated uniformity across the Mylar base (irradiation field) was 7%. Typical dose rates ranged between 0.5 and 2 Gy min^{-1}.

2.3. Sample preparation

Custom-made pyrex glass cylindrical wells were glued onto a well stretched Mylar foil (1.5 µm) by means of araldite® and cured in oven at 140 °C for 4 h. Two days prior to irradiation cells from a single-cell suspension were seeded at about 4 10^4 cells/well in 2 ml growth medium. A cell monolayer was thus obtained on the day of exposure. At the moment of irradiation the medium was aspirated and a sterile cap placed on the cylinder which was then placed horizontally in the sample holder.

For the clonogenic measurements of directly irradiated samples, cells were washed with PBS and trypsinzed from the Mylar based wells straight after the irradiation using 0.5 ml of 0.025% trypsin-EDTA incubated at 37°C for 5-10 minutes. The sample density of the resulting single cell solution was estimated by hemocytometer counting and the desired number of cells plated in T25 plastic flasks. After 14 days incubation at 37°C, 5%CO_2, colonies were fixed and stained using crystal violet solution (0.05% w/v in methanol) for manual scoring. For the bystander measurements, 2 ml medium was removed from the irradiated wells and added to cells in unexposed wells through a 0.2 μm filter straight after the irradiation. The bystander unexposed wells were then incubated at 37°C for 1 hr before following the trypsination and reseeding protocol described above.

For radiation-induced senescence experiments, cultures were propagated and tested at time intervals by means of commercially available kits (β-galactosidase assay): briefly, cells seeded in 35-mm petri dishes, fixed in paraformaldehyde and incubated in β-galactosidase (β-gal) staining solution overnight at 37 °C in the absence of CO_2. Scoring of senescent cells (positive for β-gal) was performed by means of light microscopy sampling random fields: senescent cells appeared brightly green. At least 500 cells per sample were scored.

The cell lines used for the preliminary experiments reported in this manuscript were 3 tumor cell lines (U87MG, T98G and LN229) all derived from human gliobastomas, differing slightly in their genetic make-up (i.e. tp53 status) and a normal human fibroblast cell line (AG01522). These specific cell lines were chosen as glioblastomas are one of the human tumors most suitable for hadrontherapy treatment due to their radio-resistance and difficulty in removal by surgery. On the other hand, fibroblasts are commonly found as connective and supportive tissue in many parts of the body and are inevitably exposed during external beam radiotherapy.

3. Results

3.1. Ion energy

Protons of nominal energy E = 4 MeV were accelerated by a terminal voltage V_T of 2 MV. ^{12}C ions were produced with q=5+ (VT = 2.4 MV). From measured spectra, H and ^{12}C beams emerged from Au foil with energies of 3.5 MeV and 13.7 MeV, respectively. An example of a measured spectrum after the Au foil is shown in figure 2. Further energy loss due to the Mylar sheet on which cells are seeded and, for protons, an air gap, resulted in incident energy on cells of 3.3 MeV for protons and 12.4 MeV for carbon ions.

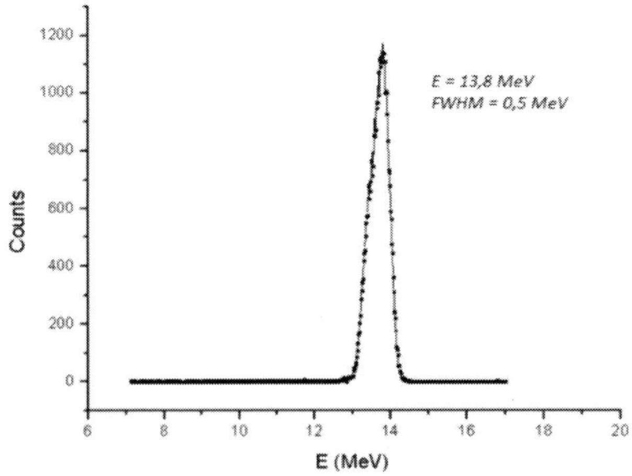

Figure 2. Measured spectrum for ^{12}C ion beam reaching the biological samples.

3.2. LET calculations

Incident energy, LET values and residual ranges were estimated using SRIM 2011 (www.srim.org) and results are reported below.

Table 1. Range of incident energies and LET values for cellular irradiation

Ion	Incident Energy [MeV]	Incident LET [keV/µm]	Residual Range in Water [µm]
p	3.3	11	174
^{12}C	12.4	730	16
^{16}O	15.5	1,100	15
^{19}F	16.7	1,350	14

As indicated in figure 2, the energy spread measured as FWHM is <5% of the mean ion energy which translates into an uncertainty in the LET calculations of ~3%. It must be noted that the LET values reported are those incident (i.e., at the surface of the biological sample).

3.3. Proton Irradiation

3.3.1. Cell survival

Survival measurements from normal human fibroblasts cells (AG01522) directly exposed to monoenergetic 3.3 MeV proton beams have been compared to data from clonogenic experiments performed with 225 kVp X-rays and 62 MeV proton from the INFN-LNS facility (figure 3). For the latter, the beam was modulated to obtain a SOBP as used for clinical applications [13] and cells exposed to the centre of the SOBP. For the X-rays and the 62 MeV proton experiments, cells were irradiated in plastic flasks but similar trypsization and re-seeding procedure were followed.

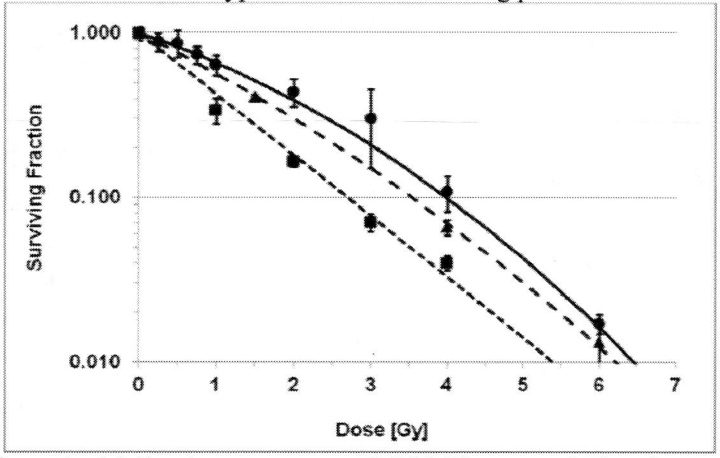

Figure 3. AG01522 survival following exposure to 3.3 MeV protons (■) and centre of SOBP for a 62 MeV proton beam (▲). 225 kVp X-ray survival curve is also reported (●).

RBE values at 10% survival ($RBE_{0.1}$) were calculated from the α and β parameters obtained fitting the experimental data with the linear quadratic model.

Table 2. RBE values for AG01522 cells irradiated with proton beams.

Proton Energy [MeV]	LET [keV/µm]	$RBE_{0.1}$
3.3	11	1.57 ± 0.07
62 - Centre SOBP	4	1.14 ± 0.02

3.3.2. Bystander effect

AG01522 cells were also used for investigation of the bystander response following 3.3 MeV proton exposures. Survival data for the bystander experiments are reported in figure 3 as a function of the dose received by the directly exposed cells from which the media was transferred.

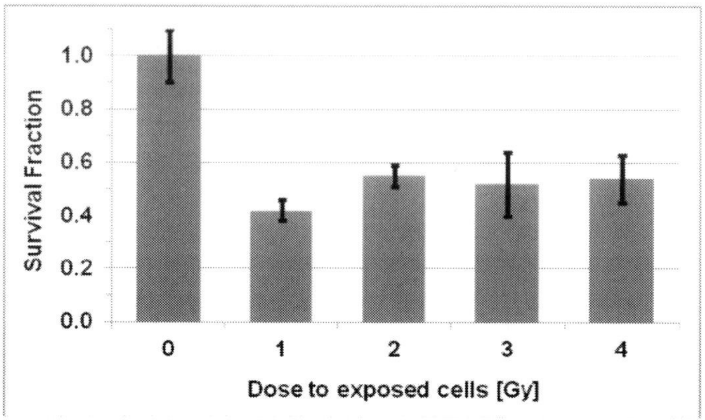

Figure 3. Survival following media transfer from AG01522 exposed to 3.3 MeV protons.

3.3.3. Senescence

The onset of cellular senescence was assayed both as an early and a late-occurring response to 3.3 MeV proton irradiation (figure 4).

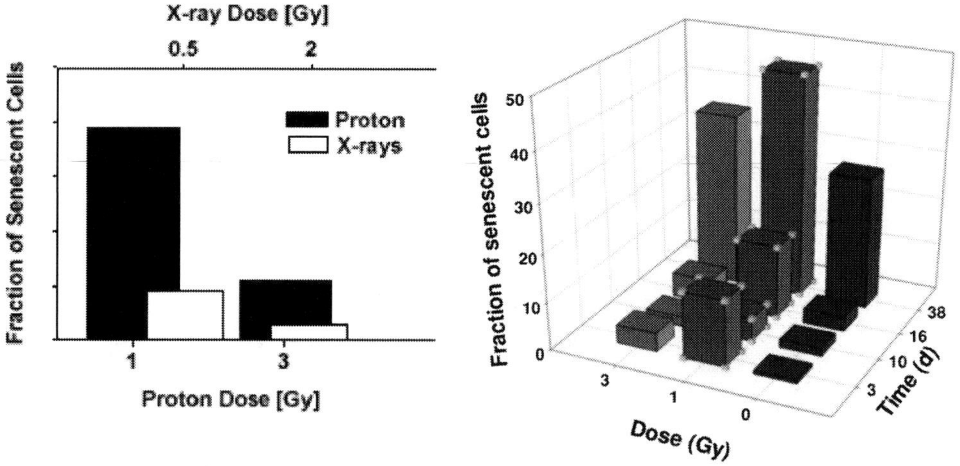

Figure 4. Early (72 hrs, left plot) and late (up to 38 days, right plot) cellular senescence post 3.3 MeV proton exposure. X-ray data are also reported for comparison.

3.4. Carbon

3.4.1. Cell survival

The biological effectiveness of proximal-to-Bragg peak (~16 μm before end of range) Carbon ion beam was investigated by measuring the survival fraction of 3 glioma cell lines (figure 5). No statistically significant differences were observed between the 3 cell lines with an estimated $RBE_{0.1}$ of ~1.7 compared to conventional X-ray beams (data not included).

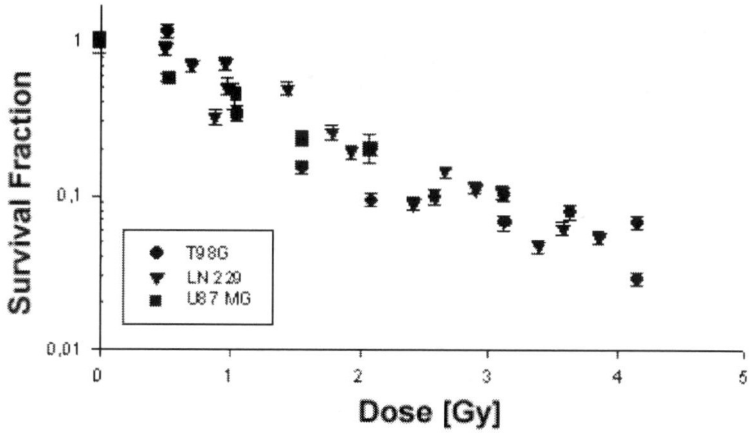

Figure 5. *Cell survival of gliobastoma cell lines exposed to 12.4 MeV ^{12}C beam*

4. Discussion

In this manuscript we report the successful upgrade of the radiobiology beam line facility at the University of Naples (Italy) with inclusion of a wide range of ion species. Carbon, Oxygen and Fluorine have been added to protons as beams that can be extracted into biological samples through a differential vacuum chamber and thin Mylar substrates. Field uniformity and dose rate are suitable for a wide range of radiation biology investigations. The maximum energy available is such to allow the heavier ions to just penetrate a single cell layer. This makes the facility ideal to investigate the biological effectiveness of the final part of the ion tracks reporting cellular effects as a function of the incident LET (as the LET varies drastically within the cell at these energies) and the atomic number (Z) of the ion beam. The effectiveness of the final part of ion tracks is critical for the optimization of RBE values to be used in clinical settings as large effects are expected and cancer cells as well as "healthy" cells may experience such traversals.

Initial biological experiments have been performed with proton and carbon beams using normal and cancer human cell lines. Survival fractions of normal human fibroblast cells (AG01522) irradiated with 3.3 MeV protons have been compared to data obtained following exposures to conventional X-ray and modulated energetic proton beams (centre of SOBP of 62 MeV protons from the INFN-LNS). Data indicate a significantly higher RBE (~1.6) for the 3.3 MeV protons compared to the centre of the SOBP (~1.1) highlighting how fixed RBE values for clinical modulated beams is an oversimplification, and considerable enhancement is to be expected in the distal part of the SOBP [14]. Interestingly, our bystander investigation shows a considerable reduction in the clonogenic potential of cells receiving media from irradiated samples. The magnitude of the effect (~50% survival reduction) could be due to the high density of cells irradiated (~10^5 cells in 2 ml) suggesting that signal concentration could modulate the bystander response. Finally, our proton data show cellular senescence significantly elevated as early as 72 hr post irradiation, contrary to what was observed with X-rays. Moreover, the level of senescent cells is constantly found elevated among proton-irradiated cultures compared to controls for up to over a month post exposure. Interestingly, at lower dose there is a higher occurrence of senescent cells which is consistent with cellular senescence as a sub-lethal stress response and in accordance with our previous results with carbon ion beams [15]. Consequences of non-lethal stress responses are of considerable interest as long term late effects, including secondary cancers, will ultimately determine the outcome of radiotherapy treatments.

Experiments with carbon beams aimed at quantifying the lethal effects of the ^{12}C track end in a panel of human glioma cell lines. No significant differences between the 3 cell lines used (U87MG, T98G and LN229) have been observed. When compared to X-rays, an RBE of ~1.7 can be estimated. This is considerably lower than what is expected in the centre of a clinical modulated carbon beam but in line with the RBE decrease at very high LETs (i.e., 700 keV/µm) [16]. In fact, according to the classical overkill effect, at such very high LET values, RBEs ought to be close to 1, hence our data may imply inadequacy of LET to describe complex effects near the Bragg peak possibly determined by the track structures. Future studies will be aimed at investigating the biological effectiveness of different types of ions with similar LET characteristics. This will provide insights into the role of delta electrons in determining the biological response to ion exposures.

References

[1] Yuh, G E, Loredo L N, Yonemoto LT, Bush D A, Shahnazi K, Preston W, Slater J M and Slater J D, Reducing toxicity from craniospinal irradiation: using proton beams to treat medulloblastoma in young children. *Cancer J.* 2004 10 386-90

[2] Kamada T, Tsujii H, Tsuji H, Yanagi T, Mizoe J E, Miyamoto T, Kato H, Yamada S, Morita S, Yoshikawa K, Kandatsu S and A. Tateishi, Efficacy and safety of carbon ion radiotherapy in bone and soft tissue sarcomas. *J. Clin. Oncol.* 2002 20 4466-71

[3] Imai R, Kamada T, Tsuji H, Yanagi T, Baba M, Miyamoto T, Kato S, Kandatsu S, Mizoe J E, Tsujii H and Tatezaki S, Carbon ion radiotherapy for unresectable sacral chordomas. *Clin. Cancer Res.* 2004 10 5741-6

[4] Jereczek-Fossa B.A., Krengli M and Orecchia R Particle beam radiotherapy for head and neck tumors: radiobiological basis and clinical experience. *Head Neck* 2006 28 750-60

[5] Jenner T J, Delara C M, O'neill P and Stevens D L, Induction and rejoining of DNA double-strand breaks in V79-4 mammalian cells following gamma and alpha irradiation. *Int. J. Radiat. Biol.* 1993 64 265-273

[6] Thacker J, Stretch A and Stephens M A, Mutation and inactivation of cultured mammalian cells exposed to beams of accelerated heavy ions. II. Chinese hamster V79 cells. *Int. J. Radiat. Biol. Relat. Stud.* Phys. Chem. Med. 1979 36 137-48

[7] Cucinotta FA, Wilson JW, Shavers M R and Katz R, Effects of track structure and cell inactivation on the calculation of heavy ion mutation rates in mammalian cells. *Int. J. Radiat. Biol.* 1996 69 593-600

[8] Durante M, K. George K, Gialanella G, Grossi G, La Tessa C, Manti L, Miller J, Pugliese M, Scampoli P and Cucinotta F A, Cytogenetic effects of high-energy iron ions: dependence on shielding thickness and material. *Radiat. Res.* 2005 164 571-6

[9] Kanai T, Endo M, Minohara S, Miyahara N, Koyama-Ito H, Tomura H, Matsufuji N, Futami Y, Fukumura A, Hiraoka T, Furusawa Y, Ando K, Suzuki M, Soga F and Kawachi K, Biophysical characteristics of HIMAC clinical irradiation system for heavy-ion radiation therapy. *Int. J. Radiat.* Oncol. Biol. Phys. 1999 44 201-10

[10] Furusawa Y, Fukutsu K, Aoki M, Itsukaichi H, Eguchi-Kasai K, Ohara H, Yatagai F, Kanai T and Ando K, Inactivation of aerobic and hypoxic cells from three different cell lines by accelerated ^{3}He-, ^{12}C- and ^{20}Ne-ion beams. *Radiat. Res.* 2000 154 485-96

[11] Uzawa A, Ando K, Koike S, Furusawa Y, Matsumoto Y, Takai N, Hirayama R, Watanabe M, Scholz M, Elsasser T and Peschke P, Comparison of biological effectiveness of carbon-ion beams in Japan and Germany. International Journal of Radiation Oncology, Biology & Physics 2009 73 1545-51

[12] Brahme A, Recent advances in light ion radiation therapy. *Int. J. Radiat. Oncol. Biol. Phys.* 2004 58 603-16

[13] Spatola C, Privitera G, Raffaele L, Salamone V, Cuttone G, Cirrone P, Sabini M G and Lo Nigro S, Clinical application of proton beams in the treatment of uveal melanoma: the first therapies carried out in Italy and preliminary results (CATANA Project). *Tumori* 2003 89 502-9

[14] Paganetti H, Niemierko A, Ancukiewicz M, Gerweck LE, Goitein M, Loeffler J S and Suit H D, Relative biological effectiveness (RBE) values for proton beam therapy. *Int. J. Radiat. Oncol. Biol. Phys.* 2002 **53** 407-21

[15] Grossi G F, Bettega D, Calzolari P, Durante M, Elsässer T, Gialanella G, Hessel P, Manti L, Pugliese M, Ritter S, Santini M T, Scampoli P and Weyrather W K, Late cellular effects of ^{12}C ions *Il Nuovo Cimento* C 2008 31 39-47

[16] Sorensen B S, Overgaard J and Bassler N, In vitro RBE-LET dependence for multiple particle types. *Acta Oncol.* 2011 **50** 757-62

AUTHOR INDEX